Math
WORKBOOK

ies Test Preparation
Integrated Educational Services

SAT®

ADVANCED PRACTICE SERIES

ILEX
Educational Publishers

Created by
Khalid Khashoggi, CEO IES
Marc Wallace

Editorial
Marc Wallace
Joseph Miller
Patrick Kennedy

Design
Kim Brown - Creative Director
Marc Wallace - Interior Designer
Ana Grigoriu, Book Cover Designer

Authors
Khalid Khashoggi, CEO IES
Marc Wallace
Joseph Miller
Meeraj Patel

Published by ILEX Publications
24 Wernik Place
Metuchen, NJ 08840
www.ILEXpublications.com
© ILEX Publications, 2014

ON BEHALF OF
Integrated Educational Services, Inc.
355 Main Street
Metuchen, NJ 08840
www.ies2400.com

We would like to thank the ILEX Publications team as well as the teachers and students at IES2400 who have contributed to the creation of this book. We would also like to thank our Chief Marketing Officer, Sonia Choi, for her invaluable input.

The SAT is a registered trademark of the College Board, which was not involved in the production of, and does not endorse, this product.

ISBN: 978-0-9913883-1-8

QUESTIONS OR COMMENTS? Email us at info@ilexpublications.com

TABLE OF CONTENTS

PRACTICE TESTS

INTRODUCTION

Welcome to the *SAT* Math Workbook* of the *IES Advanced Practice Series*. This specialized workbook is one volume in an SAT preparation series that has been developed by elite SAT teachers at Integrated Educational Services, Inc. (IES). Renowned for its easy, accurate, and efficient SAT techniques, IES has been a leader in the world of SAT preparation. Founded 15 years ago, our company is proud to have contributed to the academic and professional growth of countless students. With its innovative methods, IES is confident that this workbook will be integral to dramatically increasing math scores.

IES prides itself on delivering comprehensive SAT math techniques that can be understood and applied regardless of a student's command of math. In this IES workbook, the SAT Math has been distilled into simple-to-use techniques based on recent SAT testing and trends. Over the course of these lessons, every SAT Math problem will be taught with a demonstration example followed by ample practice. Our experience has proven that SAT Math is best approached by using simplified, clear, and straightforward problem solving methods. This book provides these methods and three full practice test.

Follow our rules, tips, strategies, and methods to a perfect score!

Dear Student:

As a test prep educator for the past 15 years, I am both pleased and honored to bring you this *SAT Math Workbook*. Here, you will find a wealth of advice on how to radically improve your mathematics score. Every year, my students absorb the techniques contained in this book, apply those techniques with care and precision, and reach their desired SAT scores. Now, I am bringing my SAT mathematics experience and knowledge directly to you, in the hope that you will enjoy similarly great successes.

In its current form, SAT mathematics is more a matter of problem solving than of anything else. This is a unique test, with a logic uniquely its own; although the principles that you learn in high school algebra and geometry are sometimes helpful, mastery of the test is much more a matter of accuracy, timing, and attention to detail. Specialized tactics and techniques that relate to these factors are what this book delivers. You will learn to pace your test-taking for maximum efficiency, and will systematically eliminate errors, as you adapt.

I encourage you to explore the full range of resources that we at ILEX and IES have made available to help you achieve your target score. This book is ideal for students who wish to give the math section special attention, but is also designed to work in tandem with the other volumes in our *Advanced Practice Series*®. Assured, informed standardized test performance is an essential first step toward a productive and deeply enjoyable college career. With these books—each one designed by expert editors and educators, each one proven to help students raise their scores—we will give you an approach to standardized testing that is straightforward, rigorous, and designed to build confidence and precision.

So get ready to take control of the SAT Mathematics test, and achieve your target score!

Sincerely,

Khalid Khashoggi
CEO, IES

TRY ALL OF OUR ADVANCED PRACTICE SERIES BOOKS

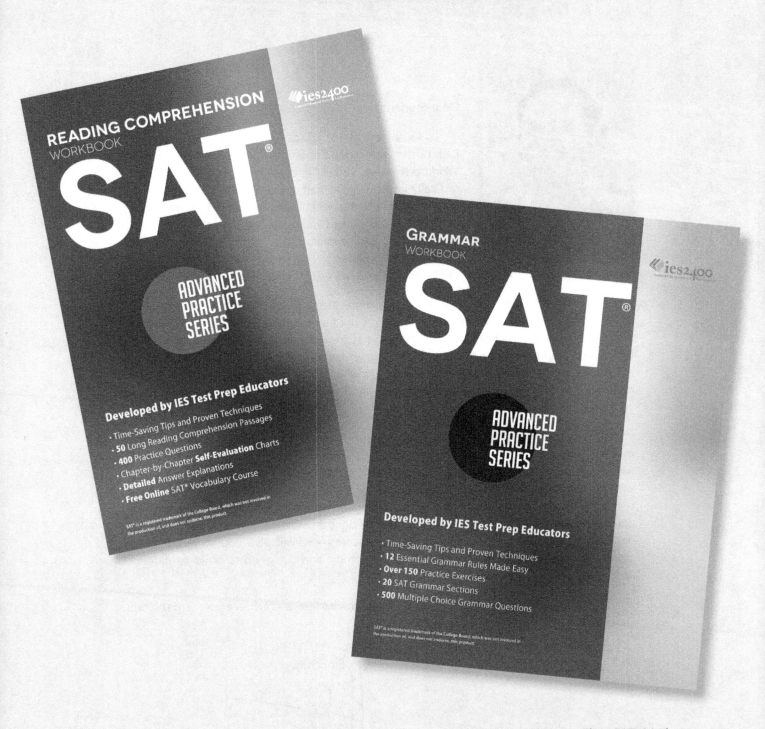

If you like this easy-to-use workbook, check out our other great volumes. The *SAT Math Workbook* is part of the *IES Advanced Practice Series* which currently includes a *Reading Comprehension Workbook*, a *Grammar Workbook*, and the soon to be released *New 2016 SAT Workbook*. Please visit www.ILEXpublications.com to order or find our complete line of SAT workbooks on Amazon.com.

Chapter 1: Word Problems

7 Sections
64 Practice Questions

1.1 Developing Word Problem Skills

Writing an equation is like translating a sentence. As you read a problem, you are translating from English to Math. Let's take a look at some common English words and phrases and how they translate into Math terms:

English	Math
product, times	Multiplication
per, ratio, proportion	Division
difference, *less than*	Subtraction
sum, *more than*	Addition
square	X^2
square root	$\sqrt{X}$

Most Importantly:
* When you see the word "is," you should put down an equal sign.
* The word "of" means to multiply. Half of x is equivalent to: $(1/2)(x)$
* "What," "a certain number," etc... All of these signify something unknown. Represent unknown quantities with a variable like **x**, or the first letter of a word in the problem. (e.g. Boys = B)

Demonstration Examples

Demo 1: 3 times a number is 5 less than 2 times that number

$$3x = 2x - 5$$

Demo 2: What is 3 more than twice itself?

$$x = 2x + 3$$

SAT Example and Technique Application:

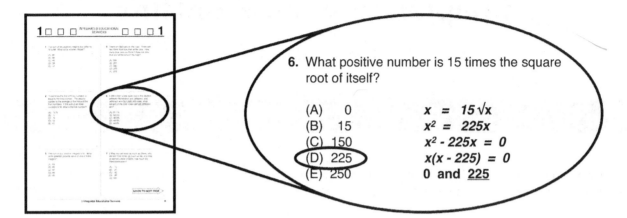

6. What positive number is 15 times the square root of itself?

(A)   0
(B)   15
(C)   150
(D)   225
(E)   250

$$x = 15\sqrt{x}$$
$$x^2 = 225x$$
$$x^2 - 225x = 0$$
$$x(x - 225) = 0$$
0 and 225

1.1 Let's Practice:

1. Ⓐ Ⓑ Ⓒ Ⓓ Ⓔ   6. Ⓐ Ⓑ Ⓒ Ⓓ Ⓔ
2. Ⓐ Ⓑ Ⓒ Ⓓ Ⓔ   7. Ⓐ Ⓑ Ⓒ Ⓓ Ⓔ
3. Ⓐ Ⓑ Ⓒ Ⓓ Ⓔ   8. Ⓐ Ⓑ Ⓒ Ⓓ Ⓔ
4. Ⓐ Ⓑ Ⓒ Ⓓ Ⓔ   9. Ⓐ Ⓑ Ⓒ Ⓓ Ⓔ
5. Ⓐ Ⓑ Ⓒ Ⓓ Ⓔ   10. Ⓐ Ⓑ Ⓒ Ⓓ Ⓔ

1. 2 less than the square root of a number is 8. What is the number?

(A) 100
(B)   36
(C)   10
(D)   6
(E) $\sqrt{10}$

2. Which equation best represents the equation defined by the following statement?

"3 less than 2 times a number is 5 less than 2 plus the number."

(A) $3 - 2x = 5 - (2 + x)$
(B) $2x - 3 = (2 + x) - 5$
(C) $3 - 2x = 5 - 2 + x$
(D) $2x - 3 = 2x - 5$
(E) $3 - 2x = 2 + x + 5$

3. When half of a number is added to the number the result is 9. What is the number?

(A) 13.5
(B) 6
(C) 9
(D) 3
(E) 4.5

4. Which equation would be best to use to find a number that is 10 more than half of itself?

(A) $x + 10 = 1/2x$
(B) $x = 1/2x - 10$
(C) $x = 2x + 10$
(D) $x + 10 = 2x$
(E) $x = 1/2x + 10$

5. A girl's age now is half her age in 6 years. What is her age in 10 years?

(A)   6
(B) 10
(C) 12
(D) 16
(E) 18

GO ON TO NEXT PAGE ⟩

1.2 Dashes

Dashes are useful in problems where **something in the problem is defined in terms of something else**.

Demonstration Examples

Demo 1: In a bag of jelly beans, there are red, green, and orange beans. There are three times as many red beans as green beans and twice as many green beans as orange beans. If there are 45 beans total in the bag, how many green jelly beans are there?

Here the number of red beans is defined in terms of green beans, which in turn are defined in terms of the orange beans. Therefore, let orange beans be represented by x, since they are smallest in number. Everything else can be defined in terms of that!

$$\underset{r}{6x} + \underset{g}{2x} + \underset{o}{x} = 45 \quad \rightarrow \quad 9x = 45 \quad \rightarrow \quad x = 5. \quad \text{Since } G = 2x, \ G = 2(5) = \underline{10}.$$

Finish completely. You may have found x, but the value of x is NOT what the question asks for.

Demo 2: The sum of three integers is 25. The second integer is twice the first, and the third integer is one more than three times the first. Find the second integer.

$$\underset{1}{x} + \underset{2}{2x} + \underset{3}{3x+1} = 25 \quad \rightarrow \quad 6x + 1 = 25 \quad \rightarrow \quad x = 4. \quad \text{So, } 2x = 2(4) = 8.$$

Demo 3: There are an equal number of lions and tigers in a zoo. When a litter of 4 lion cubs is born no no tiger cubs are born, there are twice as many lions as tigers. How many lions are there now?

$$\underline{x = \text{lions}} \quad \underline{x = \text{tigers}} \quad \rightarrow \quad \underline{x+4} = \underline{2x} \quad \rightarrow \quad x = 4. \quad \text{So, the number of lions now is } (4) + 4 = 8.$$

SAT Example and Technique Application:

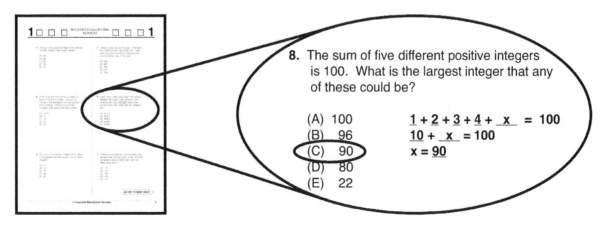

8. The sum of five different positive integers is 100. What is the largest integer that any of these could be?

(A) 100
(B) 96
(C) 90
(D) 80
(E) 22

$$1 + 2 + 3 + 4 + \underline{x} = 100$$
$$\underline{10} + \underline{x} = 100$$
$$x = \underline{90}$$

1.2 Let's Practice:

1. Ⓐ Ⓑ Ⓒ Ⓓ Ⓔ   6. Ⓐ Ⓑ Ⓒ Ⓓ Ⓔ
2. Ⓐ Ⓑ Ⓒ Ⓓ Ⓔ   7. Ⓐ Ⓑ Ⓒ Ⓓ Ⓔ
3. Ⓐ Ⓑ Ⓒ Ⓓ Ⓔ   8. Ⓐ Ⓑ Ⓒ Ⓓ Ⓔ
4. Ⓐ Ⓑ Ⓒ Ⓓ Ⓔ   9. Ⓐ Ⓑ Ⓒ Ⓓ Ⓔ
5. Ⓐ Ⓑ Ⓒ Ⓓ Ⓔ   10. Ⓐ Ⓑ Ⓒ Ⓓ Ⓔ

1. The sum of two positive integers that differ by 12 is 68. What is the smaller integer?

(A) 80
(B) 68
(C) 40
(D) 28
(E) 12

2. Three times the first of three numbers is equal to the third number. The second number is the average of the first and the third numbers. If the sum of all three numbers is 30, what is the first number?

(A) 3.75
(B) 5
(C) 6
(D) 10
(E) 15

3. The sum of four positive integers is 50. What is the greatest possible value of one of these integers?

(A) 15
(B) 40
(C) 47
(D) 48
(E) 50

4. The sum of 4 different prime numbers is 17. What is the difference between the smallest and the largest number?

(A) 3
(B) 5
(C) 7
(D) 9
(E) 11

5. There are 562 cars on the road. There are two more blue cars than white cars. How many blue cars are there if there are only blue and white cars on the road?

(A) 282
(B) 281
(C) 280
(D) 279
(E) 278

6. If 345 million votes were cast in the election between Richardson and Jefferson, and Jefferson won by 3,500,000 votes, what percent of the total votes cast did Jefferson win?

(A) 51.1%
(B) 50.5%
(C) 49.5%
(D) 48.9%
(E) 48.8%

7. If Mike earned twice as much as Steve, who earned three times as much as Ike, and they all earned a total of $200, how much did Steve alone earn?

(A) 10
(B) 20
(C) 60
(D) 80
(E) 120

GO ON TO NEXT PAGE ⟩

1.3 T-Charts

T-Charts are used when it is difficult or impossible to solve an equation algebraically. Often, questions that require T-Charts involve continuous functions, but force you to look for integer solutions. Look for the phrases "positive integers" or "integers" in the question.

Demonstration Examples

Demo 1: $x^2 - y^2 = 5$. Find x and y if they are both positive integers.

By generating a table of integer values of n and their perfect squares, you can check visually to see if there are two such perfect squares that that have a difference of 5.

n	n^2
1	1
2	4
3	9
4	16

Since 9 - 4 = 5, you can conclude that $x^2 = 9$ and $y^2 = 4$. So, x = 3 and y = 2.

Demo 2: The difference of the cube and the square of two different positive integers is 60. What is the sum of these two integers?

n	n^2	n^3
1	1	1
2	4	8
3	9	27
4	16	64

The above problem, when re-written as an equation, would be $x^3 - y^2 = 60$. This cannot be solved algebraically. Instead, use the table drawn above. After looking at the chart, you can set $x^3 = 64$ and set $y^2 = 4$. So, x = 4, y = 2, and their sum is 6.

SAT Example and Technique Application:

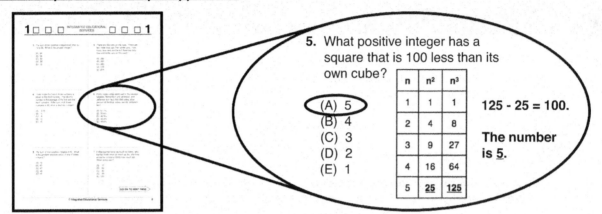

5. What positive integer has a square that is 100 less than its own cube?

(A) 5
(B) 4
(C) 3
(D) 2
(E) 1

n	n²	n³
1	1	1
2	4	8
3	9	27
4	16	64
5	**25**	**125**

125 - 25 = 100.

The number is 5.

1.3 Let's Practice:

1. Ⓐ Ⓑ Ⓒ Ⓓ Ⓔ   6. Ⓐ Ⓑ Ⓒ Ⓓ Ⓔ
2. Ⓐ Ⓑ Ⓒ Ⓓ Ⓔ   7. Ⓐ Ⓑ Ⓒ Ⓓ Ⓔ
3. Ⓐ Ⓑ Ⓒ Ⓓ Ⓔ   8. Ⓐ Ⓑ Ⓒ Ⓓ Ⓔ
4. Ⓐ Ⓑ Ⓒ Ⓓ Ⓔ   9. Ⓐ Ⓑ Ⓒ Ⓓ Ⓔ
5. Ⓐ Ⓑ Ⓒ Ⓓ Ⓔ   10. Ⓐ Ⓑ Ⓒ Ⓓ Ⓔ

1. The positive difference of the squares of two negative integers is 5. What is the product of the integers?

(A) -6
(B)   6
(C)   8
(D) -12
(E) 12

2. $X^2 - Y^2 > 5$ where X and Y are positive integers. What is the least possible value for X?

(A) 6
(B) 5
(C) 4
(D) 3
(E) 2

3. The sum of the squares of two positive integers is 52. What is the product of the integers?

(A) 12
(B) 18
(C) 24
(D) 36
(E) 48

4. The difference of the the cubes of two positive integers is 56. What is the smaller of the two integers?

(A) 1
(B) 2
(C) 3
(D) 4
(E) 5

5. The difference between the cube of a positive integer and the square of the same positive integer is four. What is the integer?

(A) 2
(B) 4
(C) 6
(D) 8
(E) 10

6. The cube of a negative number is 80 less than the square of that same number. What is the number?

(A) 4
(B) 1
(C) -1
(D) -2
(E) -4

GO ON TO NEXT PAGE ⟶

14

1.4 Aim for Your Target!

Sometimes on the SAT, you are asked to solve for something other than a variable. Because you are trained to "solve for *x*" and "simplify" in school, it may take you up to three times as long to solve for what the SAT asks you to find. Always "aim for your target" when you're answering a math question.

Demonstration Examples

Demo 1: $x + y + z = 10$ and $x + y + 2z = 6$. Find $x + y$.

Here, you could easily find out what z is, but z is NOT your target! If you look at your target you will see that you want to keep x and y. So, your goal is to eliminate z. From there, you will see that you can manipulate the equation to reach the target, x + y.

$$x + y + z = 10$$
$$x + y + 2z = 6$$

$$(x + y + z = 10)(-2)$$
$$x + y + 2z = 6$$

$$-2x - 2y - 2z = -20$$
$$\underline{x + y + 2z = 6}$$
$$-x - y = -14$$
$$\underline{\mathbf{x + y = 14}}$$

Demo 2: $(2x + 10)(2x - 10) = 20$. Find $4x^2$.

The above equation becomes: $4x^2 - 100 = 20$. This is also known as a difference of squares. Notice that you have NO NEED to find x. Just add 100 to both sides of the equation and you'll get: $4x^2 = \underline{120}$. This is your answer.

SAT Example and Technique Application:

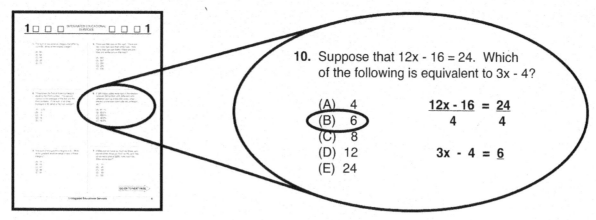

10. Suppose that $12x - 16 = 24$. Which of the following is equivalent to $3x - 4$?

(A) 4
(B) 6
(C) 8
(D) 12
(E) 24

$$\frac{12x - 16}{4} = \frac{24}{4}$$

$$3x - 4 = 6$$

1.4 **Let's Practice:**

1. Ⓐ Ⓑ Ⓒ Ⓓ Ⓔ   6. Ⓐ Ⓑ Ⓒ Ⓓ Ⓔ
2. Ⓐ Ⓑ Ⓒ Ⓓ Ⓔ   7. Ⓐ Ⓑ Ⓒ Ⓓ Ⓔ
3. Ⓐ Ⓑ Ⓒ Ⓓ Ⓔ   8. Ⓐ Ⓑ Ⓒ Ⓓ Ⓔ
4. Ⓐ Ⓑ Ⓒ Ⓓ Ⓔ   9. Ⓐ Ⓑ Ⓒ Ⓓ Ⓔ
5. Ⓐ Ⓑ Ⓒ Ⓓ Ⓔ   10. Ⓐ Ⓑ Ⓒ Ⓓ Ⓔ

1. If $x + y - z = 10$, and $x - y + z = 20$, what is x?

(A) 30
(B) 20
(C) 15
(D) 5
(E) 2

2. $2x + 3y - z = 14$ and $x + 2y = 15$. Find $y + z$.

(A) 16
(B) 14
(C) 12
(D) -14
(E) -16

3. If $(3x - 6)(3x + 6) = 20$, what is $18x^2$?

(A) 112
(B) 96
(C) 56
(D) 32
(E) 16

4. If we know that $xy = 5$, $yz = 3$, and $xz = 4$, what is the value of xyz?

(A) $\sqrt{8}$
(B) $\sqrt{60}$
(C) 8
(D) 60
(E) 120

5. If $4x + 10y = 8$, what is the value of $12x + 30y$?

(A) 72
(B) 64
(C) 32
(D) 24
(E) 12

6. Samantha bought 2 peaches and 1 grapefruit for \$4.95, and Kim bought 1 peach and two grapefruits for \$4.05. What is the combined cost of one grapefruit and one peach?

(A) \$1.50
(B) \$3.00
(C) \$4.50
(D) \$6.00
(E) \$9.00

7. If $44x = 10$, what is $22x - 3$?

(A) 7
(B) 5
(C) 3
(D) 2
(E) 1

8. Suppose that a number x is added to a number y to get a sum of 8. Also, suppose that $z - y = 6$. What is the value of $x + z$?

(A) 2
(B) 6
(C) 8
(D) 12
(E) 14

9. If x is 3 more than y, what is 4 more than y?

(A) $x - 2$
(B) $x - 1$
(C) $x + 1$
(D) $2x$
(E) x

10. The cost of one laptop and two CDs is \$426. The cost of one laptop and one CD is \$413. What is the cost of three CD's?

(A) \$13
(B) \$24
(C) \$26
(D) \$39
(E) \$42

GO ON TO NEXT PAGE ▷

1.5 Split The Difference

Demonstration Example

Demo: There are 80 more men than women in a group of 600 people. How many men and women are there?

Step 1: Get the 50% Splits.   *600 / 2 = 300 (Total Split)*   *80 / 2 = 40 (Difference Split)*

Step 2: Add Splits for greater: *(300 + 40 = 340 Men)*
Subtract Splits for lesser: *(300 - 40 = 260 Girls)*

SAT Example and Technique Application:

5. There are 18 more boys than girls in the Spanish honors society which has 42 members. How many of the members are boys?

(A) 39 $42/2 = \underline{21}$, $18/2 = \underline{9}$
(B) 30
(C) 18 $\underline{21} + \underline{9} = \underline{30}$
(D) 12
(E)   2

1.5 Let's Practice:

1. Ⓐ Ⓑ Ⓒ Ⓓ Ⓔ   6. Ⓐ Ⓑ Ⓒ Ⓓ Ⓔ
2. Ⓐ Ⓑ Ⓒ Ⓓ Ⓔ   7. Ⓐ Ⓑ Ⓒ Ⓓ Ⓔ
3. Ⓐ Ⓑ Ⓒ Ⓓ Ⓔ   8. Ⓐ Ⓑ Ⓒ Ⓓ Ⓔ
4. Ⓐ Ⓑ Ⓒ Ⓓ Ⓔ   9. Ⓐ Ⓑ Ⓒ Ⓓ Ⓔ
5. Ⓐ Ⓑ Ⓒ Ⓓ Ⓔ   10. Ⓐ Ⓑ Ⓒ Ⓓ Ⓔ

1. There are 420 more cats than dogs in an animal shelter. If the total number of cats and dogs is 1200, how many cats are there?

(A)   390
(B)   810
(C) 1020
(D) 1240
(E) 1620

2. There are 32 more boys than girls in a school of 540 students. How many girls go to the school?

(A) 236
(B) 238
(C) 254
(D) 286
(E) 302

3. Governor Smythe won the election by 2,640 votes when there were 18,714 votes cast. What percentage of the vote did he win?

(A) 42.9%
(B) 51.7%
(C) 57.0%
(D) 57.1%
(E) 64.1%

4. If there are 8,502 more nurses than doctors in a group of 88,888 doctors and nurses, how many doctors are there?

(A) 40,193
(B) 42,400
(C) 44,444
(D) 48,695
(E) 52,946

GO ON TO NEXT PAGE ⟶

1.6 Left Behind

Demonstration Example

Demo: At 8:00 am, there were x ducks in a pond. By 9:00 am, 1/4 of the ducks had flown away. At 10:00 am, 1/3 of those remaining flew away. In terms of x, how many ducks are left in the pond at 10:01 am?

What's Left? *At 9 am, 3/4 of x are remaining, and at 10 am, 2/3 of 3/4 of x are remaining. So, multiply the fractions together and get: 3/4 (2/3) (x) = x / 2.*

SAT Example and Technique Application:

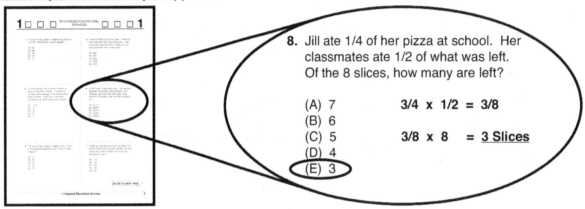

8. Jill ate 1/4 of her pizza at school. Her classmates ate 1/2 of what was left. Of the 8 slices, how many are left?

(A) 7   3/4 x 1/2 = 3/8
(B) 6
(C) 5   3/8 x 8   = 3 Slices
(D) 4
(E) 3

1.6 Let's Practice:

1. Ⓐ Ⓑ Ⓒ Ⓓ Ⓔ   6. Ⓐ Ⓑ Ⓒ Ⓓ Ⓔ
2. Ⓐ Ⓑ Ⓒ Ⓓ Ⓔ   7. Ⓐ Ⓑ Ⓒ Ⓓ Ⓔ
3. Ⓐ Ⓑ Ⓒ Ⓓ Ⓔ   8. Ⓐ Ⓑ Ⓒ Ⓓ Ⓔ
4. Ⓐ Ⓑ Ⓒ Ⓓ Ⓔ   9. Ⓐ Ⓑ Ⓒ Ⓓ Ⓔ
5. Ⓐ Ⓑ Ⓒ Ⓓ Ⓔ   10. Ⓐ Ⓑ Ⓒ Ⓓ Ⓔ

1. At Angie's birthday party, her friends were allowed first dibs on the cake. They finished 1/5 of it. Angie's family ate 1/3 of what was left over from her friends. What fraction of the original cake was Angie able to save for later?

    (A) 2/5
    (B) 7/15
    (C) 8/15
    (D) 11/15
    (E) 4/5

2. 7/8 of the Earth's surface is covered by water. 2/5 of the Earth's land mass is covered by forest. Humans currently occupy 1/3 of the land that is not covered by forest. What fraction of the Earth's surface is forest-free land that remains open for human development?

    (A) 1/20
    (B) 1/25
    (C) 1/30
    (D) 1/40
    (E) 1/50

GO ON TO NEXT PAGE ⟩

1 □ □ □   Unauthorized copying or
reuse of any part of this
page is illegal.   □ □ □ 1

1.7 Foiling and Factoring

1.7.1 It is important that you learn the following 3 expansions in the format shown below:

1. $(X + Y)^2 = X^2 + Y^2 + 2XY$
2. $(X - Y)^2 = X^2 + Y^2 - 2XY$
3. $(X + Y)(X - Y) = X^2 - Y^2$

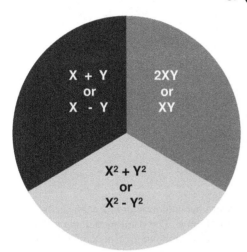

Choose Any Two Of Three!

On the SAT, a question will often give you any two of of these three pieces of information and ask you to find the third. Since the test has a tendency to group X^2 and Y^2 together, we suggest memorizing the expansions in the order shown above.

This is not an unchanging rule, just the most common style of the question. Be prepared to answer foiling and factoring questions that occur in very different arrangements. (See 1.7.2)

Demonstration Example

Demo: If $X^2 + Y^2 = 70$ and $(X - Y)^2 = 30$, find XY.

Step 1:	*Choose equation:*	$(X - Y)^2 = X^2 + Y^2 - 2XY$
Step 2:	*Substitute:*	$(30) = (70) - 2XY$
Step 3:	*Solve for XY:*	$-40 = -2XY$ So, XY = <u>20</u>.

SAT Example and Technique Application:

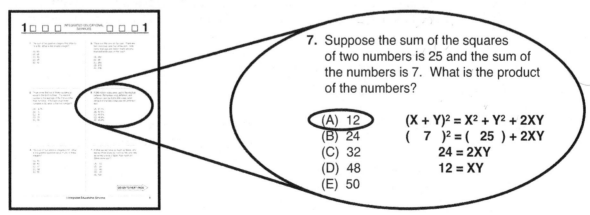

7. Suppose the sum of the squares of two numbers is 25 and the sum of the numbers is 7. What is the product of the numbers?

(A) 12
(B) 24
(C) 32
(D) 48
(E) 50

$(X + Y)^2 = X^2 + Y^2 + 2XY$
$(7)^2 = (25) + 2XY$
$24 = 2XY$
$12 = XY$

1.7.1 Let's Practice:

1. Ⓐ Ⓑ Ⓒ Ⓓ Ⓔ   6. Ⓐ Ⓑ Ⓒ Ⓓ Ⓔ
2. Ⓐ Ⓑ Ⓒ Ⓓ Ⓔ   7. Ⓐ Ⓑ Ⓒ Ⓓ Ⓔ
3. Ⓐ Ⓑ Ⓒ Ⓓ Ⓔ   8. Ⓐ Ⓑ Ⓒ Ⓓ Ⓔ
4. Ⓐ Ⓑ Ⓒ Ⓓ Ⓔ   9. Ⓐ Ⓑ Ⓒ Ⓓ Ⓔ
5. Ⓐ Ⓑ Ⓒ Ⓓ Ⓔ   10. Ⓐ Ⓑ Ⓒ Ⓓ Ⓔ

1. $x^2 - y^2 = 20$ and $x - y = 2$. What is $x + y$?

(A) 18
(B) 10
(C) 9
(D) 8
(E) 5

2. $x - y = 5$ and $xy = 24$. What is $x^2 + y^2$?

(A) 73
(B) 48
(C) 25
(D) 23
(E) 22

3. If $(x+y)^2 = 80$ and $(x - y)^2 = 16$, what is xy?

(A) 112
(B) 56
(C) 48
(D) 32
(E) 16

4. $x^2 - y^2 = 27$ and $x - y = 3$. What is x?

(A) 2
(B) 3
(C) 6
(D) 9
(E) 12

5. $x^2 - y^2 = 100$ and $x + y = 25$. What is y?

(A) 3/2
(B) 11/2
(C) 13/2
(D) 21/2
(E) 29/2

6. x and y are two integers greater than 1, and x is larger than y. The difference of the squares of x and y is 12 and the difference of x and y is 2. What is the sum of x and y?

(A) 1
(B) 2
(C) 4
(D) 6
(E) 8

1.7.2

* At other times, factoring and foiling questions require you to mirror the quadratic form of the equation that is given in order to solve for certain variables or expressions. Let's see how this looks below:

Demonstration Example

Demo: $X^2 + BX + 81 = (X + H)^2$ where B and H are constants. Find the values of B and H.

Expand the right side of the equation to get: $X^2 + BX + 81 = X^2 + 2HX + H^2$.
Now, match the corresponding "like" terms on both sides...

$X^2 = X^2$ $BX = 2HX$. So, $B = 2H$. $H^2 = 81$. So, <u>H = 9 and B = 18</u>.

1 □ □ □   Unauthorized copying or
                 reuse of any part of this
                 page is illegal.   □ □ □ 1

SAT Example and Technique Application:

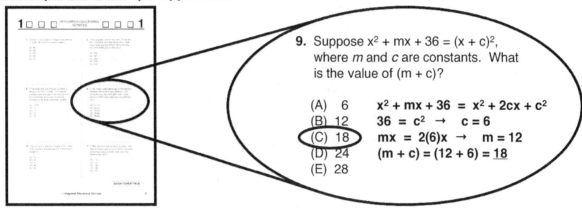

9. Suppose $x^2 + mx + 36 = (x + c)^2$, where m and c are constants. What is the value of $(m + c)$?

(A) 6 $x^2 + mx + 36 = x^2 + 2cx + c^2$
(B) 12 $36 = c^2 \rightarrow c = 6$
(C) 18 $mx = 2(6)x \rightarrow m = 12$
(D) 24 $(m + c) = (12 + 6) = \underline{18}$
(E) 28

1.7.2 **Let's Practice:**

1. Ⓐ Ⓑ Ⓒ Ⓓ Ⓔ   6. Ⓐ Ⓑ Ⓒ Ⓓ Ⓔ
2. Ⓐ Ⓑ Ⓒ Ⓓ Ⓔ   7. Ⓐ Ⓑ Ⓒ Ⓓ Ⓔ
3. Ⓐ Ⓑ Ⓒ Ⓓ Ⓔ   8. Ⓐ Ⓑ Ⓒ Ⓓ Ⓔ
4. Ⓐ Ⓑ Ⓒ Ⓓ Ⓔ   9. Ⓐ Ⓑ Ⓒ Ⓓ Ⓔ
5. Ⓐ Ⓑ Ⓒ Ⓓ Ⓔ   10. Ⓐ Ⓑ Ⓒ Ⓓ Ⓔ

1. If $4(x + y)^2 = 100$, what is the value of $x + y$?

(A) 625
(B) 100
(C) 25
(D)   5
(E)   1

2. $6x^2 + 12x + 6 = 96$. If x is positive, what is x?

(A) -3
(B) -2
(C) 3
(D) 4
(E) 5

3. Suppose that $x^2 + (z + 1)x + 81 = (x + m)^2$, where z and m are positive constants. What is z - m?

(A) 26
(B) 20
(C) 17
(D) 14
(E) 8

4. $x^2 - 14x + 8 = -41$. What is the value of x?

(A) 11
(B) 7
(C) 3
(D) 2
(E) 1

5. $(x + 1)(x + d) = x^2 + fx + 9$. What is the value of f?

(A) 10
(B) 9
(C) 8
(D) 6
(E) 5

6. $(x^2 - y^2) / (x + y) = T$, where T is a constant. What is x in terms of y and T?

(A) $T + y$
(B) $T - y$
(C) $T^2 - y$
(D) $T + y^2$
(E) $T - y^2$

GO ON TO NEXT PAGE ▷

CHAPTER 1: CHALLENGE QUESTIONS

Student-Produced Responses

1.1

1. What positive number is 15 less than the square of the sum of itself and 3?

1.2

2. The average of four positive numbers is 15. The second number is 2 times the first number. The third number is twice the second. The last number is the square of the first number. What is the largest number?

1.3

3. The difference between the cubes of two numbers is twice the value of the perfect square of a positive integer. What is the larger number minus the smaller number?

1.4

4. If $x + 3 = y - 7$, what is the value of $y^2 - 14y$ when x is equal to 10?

5. If 2 coffees and 3 sodas cost $2.64, and 4 coffees and 1 soda cost $2.78, how many dollars would it cost to buy 3 coffees and 2 sodas?

1.5

6. If 24,400,000 people voted in an election and the winner won by 1/1000 of a percent of the total vote, what is the number of votes that separate the loser from the 50th percentile?

1.6

7. Dale brought a jar of M&M's to school. His best friends ate 1/3 of the M&M's in the jar in the morning. His Spanish class ate 1/5 of what was left during class, and the principal ate 1/15 of what was left after school. If there were 336 M&M's left when Dale went home, how many were in the container to start?

8. Allison has a magical rock that breaks into thirds every time it is dropped. Every time she drops it, she loses one piece. If she drops the rock once and drops the remaining pieces again, what fraction of the original rock will she have left?

1.7

9. If $x - y = 6$ and $x^2 - y^2 = 0$, what is $|x| + |y|$?

10. Suppose that $4x^2 + 8xy + 4y^2 = 576$. If $x > 1$ and $y > 1$, what is the largest number that x could be if x is divisible by y?

GO ON TO NEXT PAGE ⟩

1 ☐ ☐ ☐   Unauthorized copying or
reuse of any part of this
page is illegal.   ☐ ☐ ☐ 1

Multiple-Choice

Student-Produced Responses

CHAPTER

1

REVIEW

1 ⓐⓑⓒⓓⓔ
2 ⓐⓑⓒⓓⓔ
3 ⓐⓑⓒⓓⓔ
4 ⓐⓑⓒⓓⓔ
ⓐⓑⓒⓓⓔ
ⓐⓑⓒⓓⓔ
ⓐⓑⓒⓓⓔ
ⓐⓑⓒⓓⓔ
ⓐⓑⓒⓓⓔ
ⓐⓑⓒⓓⓔ

5 · 6 · 7 · 8 ·

---

1. When a certain number is multiplied by five, the result is the same as the number plus five. What is the number?

   (A) 5
   (B) 4
   (C) 5/4
   (D) 4/5
   (E) 0

2. Four more than twice a number is half of the sum of that number and ten. What is the number?

   (A) 3
   (B) 2
   (C) 3/2
   (D) 2/3
   (E) 1/2

3. Max is 5 years older than his sister Hermione. In ten years he will be 3 times as old as she is now. How old will Hermione be in ten years?

   (A) 5
   (B) 7.5
   (C) 10
   (D) 12.5
   (E) 17.5

4. The cost of 2 bags of chips and 3 bottles of soda is 6 dollars. The cost of 1 bag of chips and 2 bottles of soda is 4 dollars. What is the cost of 6 bags of chips and 10 bottles of soda?

   (A) $10.00
   (B) $20.00
   (C) $30.00
   (D) $40.00
   (E) $50.00

5. A person stands 40 yards away from a wall. Each time he moves forward, he closes half the distance between him and the wall. Approximately how much distance is left between him and the wall, in feet, after the fifth time he has moved?

6. Lev, Katerina, and Stephan are selling cheap tickets to the circus. If Lev sells 20 more tickets than Stephan, who sells twice as many tickets as Katerina, how many tickets did Lev sell if the total number of tickets sold by the three of them is 120?

7. The sum of two different positive numbers is 15. The sum of their squares is 115. What is the product of the two numbers?

8. If $2d + 9 = 35$, what does $4d - 10$ equal?

STOP

Chapter 2: Percentages and Averages

4 Sections
65 Practice Questions

2.1 Percentages

2.1.1 A Percentage is a part out of a hundred. For example: 2% = 2/100 or 0.02. It is important that you get into the habit of writing percentages as decimals when working with percentage problems.

Demonstration Example

Demo: What is 30% of 70?

In school you have probably been told to set up a proportion and solve for x when doing percentage problems:

$$\frac{30}{100} = \frac{x}{70}$$

From now on, do this instead: 30% = 0.3 and the word "of" means multiply. So, 30% of 70 really means (0.3)(70) = 21. This is faster!

Suppose you wanted to know 30% of 70% of a number. 30% of 70% of x is just (0.3)(0.7)x = 0.21x. So, you really need to find 21% of x.

Use your word problem skills to your advantage. For instance, 40 is 20% of what number? Just write 40 = (0.2)(x), where the word "is" means "equals" and the word "of" means "times or multiply by". So, x = 200 in this case.

SAT Example and Technique Application:

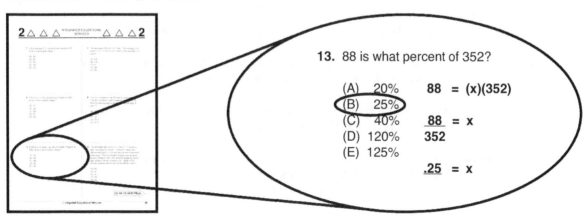

13. 88 is what percent of 352?

(A) 20%   88 = (x)(352)
(B) 25%
(C) 40% $\frac{88}{352}$ = x
(D) 120%
(E) 125%

.25 = x

2.1.1 **Let's Practice:**

1. Ⓐ Ⓑ Ⓒ Ⓓ Ⓔ   6. Ⓐ Ⓑ Ⓒ Ⓓ Ⓔ
2. Ⓐ Ⓑ Ⓒ Ⓓ Ⓔ   7. Ⓐ Ⓑ Ⓒ Ⓓ Ⓔ
3. Ⓐ Ⓑ Ⓒ Ⓓ Ⓔ   8. Ⓐ Ⓑ Ⓒ Ⓓ Ⓔ
4. Ⓐ Ⓑ Ⓒ Ⓓ Ⓔ   9. Ⓐ Ⓑ Ⓒ Ⓓ Ⓔ
5. Ⓐ Ⓑ Ⓒ Ⓓ Ⓔ   10. Ⓐ Ⓑ Ⓒ Ⓓ Ⓔ

1. What is 40% of 70?

(A) 7
(B) 14
(C) 21
(D) 28
(E) 56

2. 30 is 60% of what number?

(A) 120
(B) 60
(C) 50
(D) 40
(E) 36

3. 80 is what percent of 250?

(A) 16
(B) 32
(C) 36
(D) 40
(E) 48

4. 30% of 40 is equal to m% of 60. What is m?

(A) 20
(B) 25
(C) 30
(D) 35
(E) 40

5. If 2.5% of n is 5, what is 0.25% of n?

(A) 1
(B) 0.5
(C) 0.1
(D) 0.05
(E) 0.01

6. 4/3 of 90 is 4/5 of what number?

(A) 85
(B) 100
(C) 120
(D) 140
(E) 150

2.1.2   **PERCENT CHANGE:**   **NEW = OLD(1 ± x%)**

Demonstration Example

> **Demo:** What is the new price of a $40 shirt when the price is increased by 20%?
>
> *Use the above formula. For a percent increase, you ADD the percent to 1 and multiply: New = 40(1 + 0.2) or $48.00.*
>
> *If this problem had directed you to decrease by 20% instead, you would write 1 MINUS 20%:   New = Old(1 - 0.2)*
>
> *In other words, increasing by 20% is the same as multiplying by 1.2, and decreasing by 20% is the same as multiplying by 0.8.*

SAT Example and Technique Application:

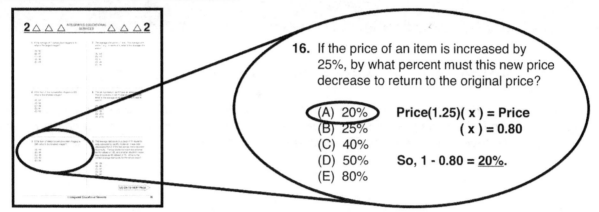

16. If the price of an item is increased by 25%, by what percent must this new price decrease to return to the original price?

(A) 20%
(B) 25%
(C) 40%
(D) 50%
(E) 80%

Price(1.25)(x) = Price
(x) = 0.80

So, 1 - 0.80 = 20%.

2.1.2 Let's Practice:

1. Ⓐ Ⓑ Ⓒ Ⓓ Ⓔ   6. Ⓐ Ⓑ Ⓒ Ⓓ Ⓔ
2. Ⓐ Ⓑ Ⓒ Ⓓ Ⓔ   7. Ⓐ Ⓑ Ⓒ Ⓓ Ⓔ
3. Ⓐ Ⓑ Ⓒ Ⓓ Ⓔ   8. Ⓐ Ⓑ Ⓒ Ⓓ Ⓔ
4. Ⓐ Ⓑ Ⓒ Ⓓ Ⓔ   9. Ⓐ Ⓑ Ⓒ Ⓓ Ⓔ
5. Ⓐ Ⓑ Ⓒ Ⓓ Ⓔ   10. Ⓐ Ⓑ Ⓒ Ⓓ Ⓔ

1. A 15% tip is added to a customer's check of $80. What is the total price that the customer pays?

(A) $81.50
(B) $85.00
(C) $88.00
(D) $92.00
(E) $95.00

2. What is the base value of a meal if, after adding a 20% tip, the total check comes to $60?

(A) $45.00
(B) $50.00
(C) $52.00
(D) $55.00
(E) $72.00

3. If a bank pays 12% interest on your savings and you deposit $2000 what is your balance in one year?

(A) $2,240.00
(B) $2,140.00
(C) $1,200.00
(D) $240.00
(E) $120.00

4. If you invest $200 in an account and your balance is $208 after one year, what is the interest rate?

(A) 1%
(B) 2%
(C) 4%
(D) 8%
(E) 12%

5. If the sales tax is 6% and you want to spend $530, what is the amount you that you can spend on goods before the tax is applied?

(A) $432.00
(B) $450.00
(C) $461.80
(D) $498.20
(E) $500.00

6. What is 240 decreased by 60%?

(A) 72
(B) 84
(C) 96
(D) 120
(E) 144

7. What is the sale price of a $40 shirt that is being sold at a 20% discount?

(A) $32.00
(B) $30.00
(C) $28.00
(D) $24.00
(E) $20.00

GO ON TO NEXT PAGE ▷

2.2 Combined Percentages

FORMULA 1:   NEW = OLD$(1 ± A\%)(1 ± B\%)(1 ± C\%)$

You have already used this equation in the previous section. But if you are applying multiple percent changes to a quantity, you may continue by simply multiplying them. For example, if an item costs $100, and its price is increased by 50% and then decreased by 20%, the new price is just:

New = 100(1.5)(0.8) = 120

FORMULA 2: $1 ± x\% = (1 ± A\%)(1 ± B\%)(1 ± C\%)$

Consider this: If a 20% increase is followed by a 40% increase, what is the overall percent increase?

1 + x% = (1.2)(1.4) = 1.68 or a 68% increase

Consider this: If the width of a door is increased by 30%, and the length of the door is increased by 20%, the new area of the door is what percent of the old area?

1 + x% = (1.3)(1.2) = 1.56 or 156% of the old area

Consider this: If the width of the above door is increased by 30%, and the length of the door is increased by 20%, by "what percent has the area increased overall"?

Well, if 1 + x% = 1.56, the overall increase is 56%.

SAT Example and Technique Application:

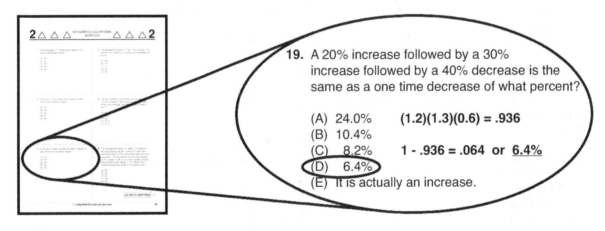

19. A 20% increase followed by a 30% increase followed by a 40% decrease is the same as a one time decrease of what percent?

(A) 24.0%   (1.2)(1.3)(0.6) = .936
(B) 10.4%
(C) 8.2%   1 - .936 = .064 or 6.4%
(D) 6.4%
(E) It is actually an increase.

2.2 Let's Practice:

1. Ⓐ Ⓑ Ⓒ Ⓓ Ⓔ   6. Ⓐ Ⓑ Ⓒ Ⓓ Ⓔ
2. Ⓐ Ⓑ Ⓒ Ⓓ Ⓔ   7. Ⓐ Ⓑ Ⓒ Ⓓ Ⓔ
3. Ⓐ Ⓑ Ⓒ Ⓓ Ⓔ   8. Ⓐ Ⓑ Ⓒ Ⓓ Ⓔ
4. Ⓐ Ⓑ Ⓒ Ⓓ Ⓔ   9. Ⓐ Ⓑ Ⓒ Ⓓ Ⓔ
5. Ⓐ Ⓑ Ⓒ Ⓓ Ⓔ   10. Ⓐ Ⓑ Ⓒ Ⓓ Ⓔ

1. A 20% increase followed by a 30% increase is equivalent to a one time increase of what %?

(A) 56%
(B) 58%
(C) 60%
(D) 150%
(E) 156%

2. A 40% increase followed by a 60% decrease is equivalent to a one time decrease of what %?

(A) 20%
(B) 44%
(C) 50%
(D) 56%
(E) 60%

3. The length of a rectangle is increased by 40% and the width is increased by 20%. The new area is what percent of the old area?

(A)   8%
(B) 56%
(C) 68%
(D) 108%
(E) 168%

4. The sides of a square are increased by 20%. By what % did the area increase?

(A) 24%
(B) 40%
(C) 44%
(D) 140%
(E) 144%

5. The sides of a cube are increased by 20%. The new volume is what percent of the old volume?

(A) 44.0%
(B) 72.8%
(C) 80.0%
(D) 144.0%
(E) 172.8%

6. If the temperature starts at 75^0, increases by 12%, then decreases by 15%, what will the new temperature be?

(A) 70.0°
(B) 71.4°
(C) 72.0°
(D) 76.4°
(E) 77.0°

7. If X is 80% of Y and Y is 150% of Z, then X is what percent of Z?

(A) 20%
(B) 70%
(C) 120%
(D) 230%
(E) 330%

8. Pacey's is having a 35% off sale. The tax rate is 6%. Jenny purchases a jacket for which she has a 10% off coupon and, after tax and all discounts, the total payment is $124.02. What was the original price of the jacket?

(A) $100
(B) $130
(C) $180
(D) $200
(E) $230

9. If the average wage at a company starts at $12/hr. and increases by 15% the first year, goes up by 6% the second year, up by 2% the third year, and finally decreases by 10% the fourth year, what is the average wage after 4 years?

(A) $11.68
(B) $12.24
(C) $12.42
(D) $13.43
(E) $13.56

GO ON TO NEXT PAGE ▷

2.3 Averages

Definitions:

Mean = Sum ÷ Number
Median: Middle Number (Located in a series of n numbers at (n+1)/2)
Mode: Most frequently occurring number

For groups of evenly spaced numbers, such as these below, ***the mean and the median are equal.***

1, 2, 3, 4, 5, ...	These are known as consecutive numbers.
2, 4, 6, 8, 10, ...	These are known as consecutive even numbers.
1, 3, 5, 7, 9, ...	These are known as consecutive odd numbers.
5, 10, 15, 20, 25, ...	These are numbers that are evenly spaced by 5.

Notice that the median in each of these cases is equal to the average of each pair of numbers that are equidistant from the first and last terms in the sequence. For instance, suppose in the fourth sequence above that the last term is 25. The average of 5 and 25 and the average of 10 and 20 are both equal to 15. This is true even when there is an even number of numbers in the list, with no number exactly in the middle. Look at the sequence 2, 4, 6, 8, 10, 12. The "median" would be situated exactly half way between 6 and 8 and is equal to (6 + 8) / 2, which is 7.

Demonstration Examples

Demo 1: For which of the following is the mean greater than the median?

(A) 2, 3, 4, 5, 6, 7, 8
(B) 2, 3, 3, 5, 6, 7, 8
(C) 2, 3, 4, 5, 6, 7, 7
(D) 2, 3, 4, 5, 6, 7, 9
(E) 5, 5, 5, 5, 5, 5, 5

In answer choice A, the numbers are consecutive, so the mean = the median, which is 5. The mean is the sum of all the terms divided by how many there are. So, if you reduce the value of any of the numbers in the list, you are lowering the sum, which lowers the mean. Consequently, if you increase any of the numbers in the list, you are increasing the sum, which increases the mean. The answer here is D, since 8 has been increased to 9.

Demo 2: The average of 5 consecutive numbers is 72. What is the greatest of these numbers?

Since the mean and the median of consecutive numbers are the same thing, we can write:

____ ____ 72 ____ ____ , *with 72 as the median.*

Since the numbers are consecutive, we can fill in the blanks and count up to the greatest integer.

____ ____ 72 73 74   *So, 74 is the greatest number.*

2 △ △ △ △ △ △ 2

Unauthorized copying or
reuse of any part of this
page is illegal.

More Demonstration Examples

Demo 3: In a list of 51 consecutive integers, the median is 80. What is the greatest integer?

For big lists, find the distance from the median to either end using (n-1) / 2 and add to or subtract from the median to find the greatest or least number.

$$(51 - 1) / 2 = 25$$
$$So, 80 + 25 = \underline{105}!$$

Demo 4: The average of x and y is 6. The average of x, y, w, and z is 24. What is the average of w and z?

This is possibly one of the SAT's all time favorite question types -- writing and manipulating equations that involve averages. Let's write equations using the information in the question:

$$\frac{x+y}{2} = 6 \qquad \frac{x+y+w+z}{4} = 24$$

From these equations, you can figure out that x + y = 12 and x + y + w + z = 96. By simple substitution, you discover that w + z = 84. So,

$$\frac{w+z}{2} = \frac{84}{2} = \underline{42}$$

SAT Example and Technique Application:

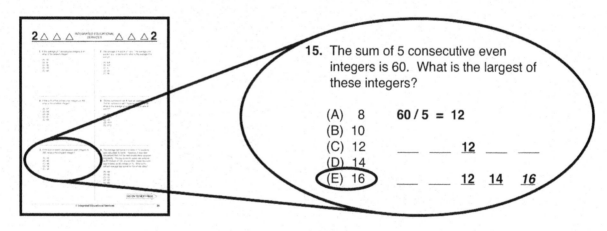

15. The sum of 5 consecutive even integers is 60. What is the largest of these integers?

(A) 8   60 / 5 = 12
(B) 10
(C) 12   ___ ___ 12 ___ ___
(D) 14
(E) 16   ___ ___ 12 14 16

2.3 **Let's Practice:**

1. Ⓐ Ⓑ Ⓒ Ⓓ Ⓔ   6. Ⓐ Ⓑ Ⓒ Ⓓ Ⓔ
2. Ⓐ Ⓑ Ⓒ Ⓓ Ⓔ   7. Ⓐ Ⓑ Ⓒ Ⓓ Ⓔ
3. Ⓐ Ⓑ Ⓒ Ⓓ Ⓔ   8. Ⓐ Ⓑ Ⓒ Ⓓ Ⓔ
4. Ⓐ Ⓑ Ⓒ Ⓓ Ⓔ   9. Ⓐ Ⓑ Ⓒ Ⓓ Ⓔ
5. Ⓐ Ⓑ Ⓒ Ⓓ Ⓔ   10. Ⓐ Ⓑ Ⓒ Ⓓ Ⓔ

1. If the average of 7 consecutive integers is 41, what is the largest integer?

   (A) 40
   (B) 41
   (C) 42
   (D) 43
   (E) 44

2. If the sum of five consecutive integers is 305, what is the smallest integer?

   (A) 57
   (B) 58
   (C) 59
   (D) 61
   (E) 63

3. If the sum of seven consecutive even integers is 350, what is the smallest integer?

   (A) 44
   (B) 46
   (C) 48
   (D) 52
   (E) 56

4. If the sum of six consecutive even integers is 186, what is the smallest integer?

   (A) 24
   (B) 26
   (C) 30
   (D) 32
   (E) 36

5. If the average of x and y is 3, and the average of x, y, and z is 12, then what is the value of z?

   (A) 6
   (B) 12
   (C) 24
   (D) 30
   (E) 36

6. If the sum of n consecutive integers is 42, then **n** could be:

   I.   4
   II. 6
   III. 7

   (A) I only
   (B) II only
   (C) III only
   (D) I and III
   (E) I, II, and III

7. The average of k and k + 1 is x. The average of k and k-1 is y. In terms of k, what is the average of x and y?

   (A) k/4
   (B) k/2
   (C) k
   (D) 2k
   (E) 4k

8. The ten numbers in set X have an average of 30. The 30 numbers in set Y have an average of 20. What is the average of all 40 numbers in sets X and Y?

   (A) 22
   (B) 22.5
   (C) 25
   (D) 25.5
   (E) 27.5

9. The average test score in a class of 10 students was calculated to be 90. However, it was later discovered that 2 of the test scores were recorded incorrectly. The top student's score was entered as 95 instead of 100, and another student's score was entered as 90 instead of 75. What is the correct average test score for the whole class?

   (A) 89
   (B) 90
   (C) 91
   (D) 93
   (E) 95

GO ON TO NEXT PAGE ▷

2.4 Charts and Tables

The SAT often tests your understanding of percentages and averages by asking you to interpret charts and tables. These charts and tables can include frequency tables, pie charts, bar graphs, and other organizational displays of data.

Demonstration Examples

Demo 1:   Find the Mean, Mode, and Median *number of pets per household* in the table below.

Households	Number of Pets	Total Pets
1	0	1x0 = 0
10	1	10x1 = 10
5	2	5x2 = 10
6	3	6x3 = 18
2	4	2x4 = 8
1	5	1x5 = 5
Total Households: 25		Total Pets: 51

Mean: *Total Pets ÷ Total Households = 51 ÷ 25 ≈ 2*

Mode: *1 (There are 10 homes with one pet, making 1 the most frequently occurring number.)*

Median: *There are 25 homes. The median number of pets will be the number of pets in the median home: (25 + 1) ÷ 2 = 13th home. The first home has zero pets, the next ten homes have 1 pet, so, the 13th home falls within the next 5 homes and has 2 pets. The median number of pets is 2.*

Demo 2:   According to the relative frequency pie chart, what are the answers to the following questions?

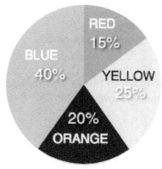

**FAVORITE COLORS OF
STUDENTS IN THE
SOPHOMORE CLASS**

1. What is the most popular color?
Just look for the largest sector -- Blue!

2. If 80 students answered red or yellow, how many students are there in the sophomore class? **Well, 40% of the students chose red or yellow and 80 sophomores answered red or yellow. So, .40(x) = 80. So, 200.**

3. Given the answer to question 2, how many students like orange? **If there are 200 students as we found out in question 2, then according to the pie chart 20% of them prefer orange. So, (0.2)(200) = 40 kids.**

2.4 Let's Practice:

1. Ⓐ Ⓑ Ⓒ Ⓓ Ⓔ   6. Ⓐ Ⓑ Ⓒ Ⓓ Ⓔ
2. Ⓐ Ⓑ Ⓒ Ⓓ Ⓔ   7. Ⓐ Ⓑ Ⓒ Ⓓ Ⓔ
3. Ⓐ Ⓑ Ⓒ Ⓓ Ⓔ   8. Ⓐ Ⓑ Ⓒ Ⓓ Ⓔ
4. Ⓐ Ⓑ Ⓒ Ⓓ Ⓔ   9. Ⓐ Ⓑ Ⓒ Ⓓ Ⓔ
5. Ⓐ Ⓑ Ⓒ Ⓓ Ⓔ   10. Ⓐ Ⓑ Ⓒ Ⓓ Ⓔ

Use the chart below to answer questions 1 to 3.

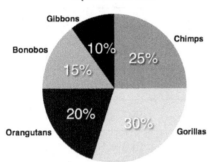

Apes at the zoo

1. If the zoo has 180 gorillas, what is the total number of apes at the zoo?

   (A) 180
   (B) 500
   (C) 520
   (D) 540
   (E) 600

2. If there are 200 apes in the zoo, the number of orangutans is greater than the number of bonobos by how much?

   (A) 5
   (B) 10
   (C) 15
   (D) 30
   (E) 100

3. Given the answer to question 1, how many gibbons and bonobos together are there in the zoo?

   (A) 100
   (B) 120
   (C) 150
   (D) 200
   (E) 300

Use the chart below to answer questions 4 to 7.

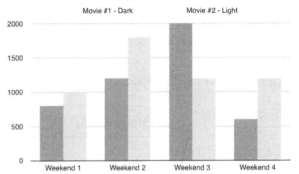

Weekend Earnings of 2 Movies Over 4 Weeks

Movie #1 - Dark   Movie #2 - Light

4. In which weekend was the ratio of earnings for Movie 2 to Movie 1 the greatest?

   (A) Weekend 1
   (B) Weekend 2
   (C) Weekend 3
   (D) Weekend 4
   (E) It cannot be determined.

5. After which weekend do you see the greatest change in earnings for either Movie?

   (A) Weekend 1
   (B) Weekend 2
   (C) Weekend 3
   (D) Weekend 4
   (E) It cannot be determined.

6. For Movie 1, which weekend had the greatest increase in earnings from the previous week?

   (A) Weekend 1
   (B) Weekend 2
   (C) Weekend 3
   (D) Weekend 4
   (E) It cannot be determined.

7. Between which two weekends was the percentage change for viewers of Movie 2 the smallest?

   (A) 3 to 4
   (B) 2 to 4
   (C) 1 to 4
   (D) 2 to 3
   (E) 1 to 3

GO ON TO NEXT PAGE ⇨

2 △ △ △ △ △ △ 2

Unauthorized copying or
reuse of any part of this
page is illegal.

CHAPTER 2: CHALLENGE QUESTIONS

Student-Produced Responses

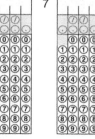

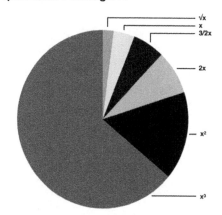

2.1

1. Jane and Alex have the same amount of trading cards. Jane gives Alex half of her playing cards. If Jane discards 25% of her trading cards, Alex discards 50% of his trading cards, and afterward they have a total of 90 cards, how many cards did they each possess in the beginning?

2.2

2. Bill invested $422.33 in a friend's business. The profits from the business increased his investment by the same percentage 3 weeks in a row. After the three weeks had passed, his investment was worth $488.90. What percent of his initial investment did Bill have after 1 week had passed? (Round to the nearest percent.)

3. After an increase, a certain number is x percent of its original value. After a decrease, that certain number is x/2 percent of its original value. If the number receives the increase followed by the decrease, the number will not change at all. What is the value of x? (Round to the nearest whole number.)

2.3

4. The average of three numbers, x, y, and z, is the same as the average of the numbers x, y, z, M, and P. The sum of x, y, and z is what percent larger than the sum of M and P?

2.4

Use the following pie chart to answer questions 5 through 7.

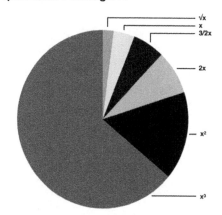

5. If x is a positive integer and the expressions in the legend above the pie chart represent the percentages in each sector, what is the percentage represented by the largest sector?

6. If the largest sector represents 224 people, how many people are represented by the smallest sector?

7. Suppose this pie chart represents a dart board. If you are going to randomly throw a dart at the dartboard and you know that the dart will definitely make contact with the board, what is the probability that the dart will hit a sector whose percentage is a multiple of 8? (Answer in decimal form.)

GO ON TO NEXT PAGE ⟩

| Multiple-Choice | Student-Produced Responses |

CHAPTER 2 REVIEW

1 Ⓐ Ⓑ Ⓒ Ⓓ Ⓔ
2 Ⓐ Ⓑ Ⓒ Ⓓ Ⓔ
3 Ⓐ Ⓑ Ⓒ Ⓓ Ⓔ
4 Ⓐ Ⓑ Ⓒ Ⓓ Ⓔ
5 Ⓐ Ⓑ Ⓒ Ⓓ Ⓔ
Ⓐ Ⓑ Ⓒ Ⓓ Ⓔ
Ⓐ Ⓑ Ⓒ Ⓓ Ⓔ
Ⓐ Ⓑ Ⓒ Ⓓ Ⓔ
Ⓐ Ⓑ Ⓒ Ⓓ Ⓔ
Ⓐ Ⓑ Ⓒ Ⓓ Ⓔ

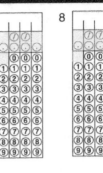

1. 90 is what percent of 300?

(A) 30
(B) 25
(C) 20
(D) 15
(E) 10

2. If k percent of 200 is 60, what is the value of k?

(A) 33
(B) 30
(C) 20
(D) .60
(E) .30

3. 30% of 40% is what percent of 60%?

(A) 10
(B) 15
(C) 20
(D) 25
(E) 30

4. 80% of 90 is what percent of 72?

(A) .01
(B) .1
(C) 1
(D) 10
(E) 100

5. A turkey costs $40. Its price is raised by 90% and then decreased by 50%. What is the final selling price of the turkey?

(A) $76
(B) $60
(C) $46
(D) $38
(E) $20

6. A tank at the aquarium is a rectangular prism. The width is increased by 20%, the length is increased by 25%, and the height is increased by 40%. The new volume of the tank is what percent of the old volume?

7. The sum of 11 consecutive even integers is 440. What is the largest of these integers?

8. The arithmetic mean of 8 consecutive even integers is 41. What is the value of the smallest integer in the group?

9. The average of L, M, N, and O is 40. The average of L, M, N, O, and P is 66. What is the value of P?

10. In mitosis, one cycle of cell division creates two cells from a single cell in 30 minutes. If 10 cells are left in a petri dish for 4.5 hours, what is the final count of cells in the dish?

GO ON TO NEXT PAGE ⟶

Multiple-Choice   **Student-Produced Responses**

CHAPTER
1 & 2
CUMULATIVE
REVIEW

1 Ⓐ Ⓑ Ⓒ Ⓓ Ⓔ
2 Ⓐ Ⓑ Ⓒ Ⓓ Ⓔ
3 Ⓐ Ⓑ Ⓒ Ⓓ Ⓔ
4 Ⓐ Ⓑ Ⓒ Ⓓ Ⓔ
5 Ⓐ Ⓑ Ⓒ Ⓓ Ⓔ
Ⓐ Ⓑ Ⓒ Ⓓ Ⓔ
Ⓐ Ⓑ Ⓒ Ⓓ Ⓔ
Ⓐ Ⓑ Ⓒ Ⓓ Ⓔ
Ⓐ Ⓑ Ⓒ Ⓓ Ⓔ
Ⓐ Ⓑ Ⓒ Ⓓ Ⓔ

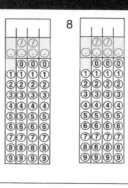

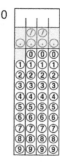

---

1. If $5n + 20 = 125$, what is the value of $15n$?

(A) 21
(B) 105
(C) 108
(D) 315
(E) 525

2. If $2a + b = 11$ and $4a^2 + b^2 = 101$, what is the value of ab?

(A) 4
(B) 5
(C) 6
(D) 12
(E) 20

3. The average of 22, 13, and what number is 15?

(A) 27
(B) 25
(C) 20
(D) 15
(E) 10

4. The arithmetic mean of 61 consecutive integers is 50. What is the greatest integer in the group?

(A) -11
(B) 20
(C) 50
(D) 70
(E) 80

5. M is 80% of N, which is 30% of P. M is what percent of P?

(A) 24.0
(B) 30.0
(C) 32.5
(D) 37.5
(E) 40.0

6. The price of a stock rose by 30 percent then dropped by 30 percent. What percent of the old price is the new price of the stock?

7. Harold earned 40 more college credits at a county college than he did when he studied abroad in Spain. A third of the number of credits he earned at the county college is 3 times as many credits as he earned while studying abroad. How many credits did he earn altogether?

8. During a rescue operation in a mine shaft, a rectangular box of supplies needs to be lowered into the ground. In order for the crate to fit through, a square opening must have its length increased by at least 30% and its width increased by at least 40%. By what percent does the area of the opening need to be increased in order for the supplies to be lowered into the mine shaft?

9. If $X + Y + Z = 110$ and $X - Y + Z = 90$, what is the average of X and Z?

10. When 8 is added to a number, the result is the same as when that number is divided by 4 and then multiplied by 5. What is the number?

STOP

Chapter 3: Ratios and Sequences

8 Sections
97 Practice Questions

3.1 Direct & Inverse Variation

3.1.1 Direct Variation: Y/X = K

A **ratio** is a way of relating quantities through a quotient. Let's say that in a particular class, the number of boys is twice the number of girls: B/G = 2/1. Sometimes, ratios are expressed using a colon: B:G = 2:1.

A **direct variation** is just another name for a **ratio** between 2 quantities Y and X, where Y/X = K. No matter what the values of Y and X are, they have a ratio that is equal to a fixed number or **constant** K.

The SAT will NOT always explicitly tell you that you are dealing with direct variation or a ratio. You will have to be able to recognize the familiar wording that goes with these questions in order to know that you are dealing with a ratio problem. Once you know that 2 quantities are related by a ratio, you can set up a **proportion** to solve the problem. Consider the example below:

Demonstration Example

Demo: For every 5 gallons of gas in an RV, or recreational vehicle, the RV can travel 36 miles. How far can the vehicle go if it has 2.5 gallons in its tank?

"For every 5 gallons...36 miles." --> This kind of phrasing should make you think of setting up a proportion immediately. So, gallons, G, and miles, M, are related by ratio:

$$\frac{G}{M} = \frac{5}{36}$$

This is our direct variation or ratio. Now we can set up a proportion and answer the question.

$$\frac{G}{M} = \frac{5}{36} = \frac{2.5}{M}$$

You can simply cross-multiply and solve for M, or you might notice that since 2.5 is half of 5, the answer must be half of 36, or 18 miles.

3   **3**   **3**   **3**   Unauthorized copying or
reuse of any part of this
page is illegal.   **3**   **3**   **3**   **3**

SAT Example and Technique Application:

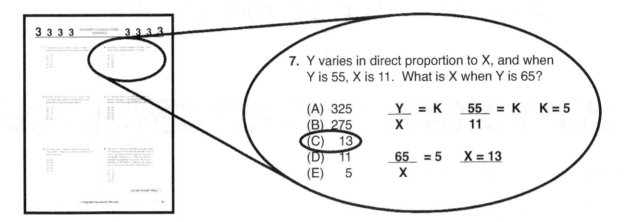

7. Y varies in direct proportion to X, and when Y is 55, X is 11. What is X when Y is 65?

(A) 325 $\frac{Y}{X} = K$ $\frac{55}{11} = K$ $K = 5$
(B) 275
(C) 13
(D) 11 $\frac{65}{X} = 5$ $X = 13$
(E) 5

3.1.1 **Let's Practice:**

1. Ⓐ Ⓑ Ⓒ Ⓓ Ⓔ   6. Ⓐ Ⓑ Ⓒ Ⓓ Ⓔ
2. Ⓐ Ⓑ Ⓒ Ⓓ Ⓔ   7. Ⓐ Ⓑ Ⓒ Ⓓ Ⓔ
3. Ⓐ Ⓑ Ⓒ Ⓓ Ⓔ   8. Ⓐ Ⓑ Ⓒ Ⓓ Ⓔ
4. Ⓐ Ⓑ Ⓒ Ⓓ Ⓔ   9. Ⓐ Ⓑ Ⓒ Ⓓ Ⓔ
5. Ⓐ Ⓑ Ⓒ Ⓓ Ⓔ   10. Ⓐ Ⓑ Ⓒ Ⓓ Ⓔ

1. Y varies in direct relation to X. When Y is 8, X is 2. What is Y when X is 7?

(A) 1.75
(B) 3
(C) 4
(D) 28
(E) 56

2. y + 64 is directly proportional to the square root of x. If y is 36 when x is 100, what is y when x is 25?

(A) -15
(B) -14
(C) 0
(D) 14
(E) 15

3. For every 100 inches of rainfall per year, a given tree in the rainforest will grow 8 ft. How much will such a tree grow if there are 250 inches of rain in a year?

(A) 20 ft.
(B) 25 ft.
(C) 30 ft.
(D) 40 ft.
(E) 50 ft.

4. Y varies directly in relation to the square of X. When Y is 45, X is 3. What is X when Y is 320?

(A) 2
(B) 4
(C) 8
(D) 12
(E) 16

GO ON TO NEXT PAGE ▷

3.1.2 Inverse Variation: $YX = K$ or $Y_1X_1 = Y_2X_2$

An inverse variation is NOT a ratio. However, inverse variations are often discussed along with direct variations/ratios, so we will mention them here. In **inverse (or indirect)** relationships, one quantity decreases in value as the other grows in value.

One example of an inverse variation is the relationship between air pressure and volume in a balloon. If you squeeze a balloon, you will decrease its volume (the amount of space it takes up), but additional air pressure will build up inside. Therefore, the two variables, pressure and volume, change in opposite directions. This is just one example of how an inverse relationship works. You do NOT need to bring knowledge from your science classes to your SAT prep. You will either be told that there is an inverse relationship, or it will be possible to identify this kind of relationship using common sense.

Demonstration Example

Demo: Y varies inversely in relation to X. If Y is 10 when X is 5, what is Y when X is 2?

In school, perhaps you were taught to first find the constant K by plugging in values for Y and X. Do so here, then plug K back into your equation. Then use your new equation to find Y when X equals something else.

$YX = K$ $(10)(5) = K$ $50 = K$ $YX = 50$ $Y(2) = 50$ $\underline{Y = 25}$.

From now on, just do this: (A faster way!)

$Y_1X_1 = Y_2X_2$ $(10)(5) = (Y)(2)$ *It is easy to see that $\underline{Y = 25}$ when $\underline{X = 2}$.*

SAT Example and Technique Application:

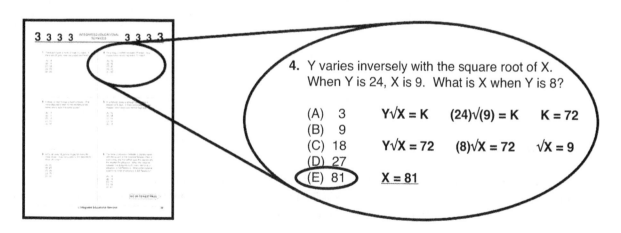

4. Y varies inversely with the square root of X. When Y is 24, X is 9. What is X when Y is 8?

(A) 3
(B) 9
(C) 18
(D) 27
(E) 81

$Y\sqrt{X} = K$ $(24)\sqrt{(9)} = K$ $K = 72$

$Y\sqrt{X} = 72$ $(8)\sqrt{X} = 72$ $\sqrt{X} = 9$

$\underline{X = 81}$

3.1.2 Let's Practice:

1. Ⓐ Ⓑ Ⓒ Ⓓ Ⓔ 6. Ⓐ Ⓑ Ⓒ Ⓓ Ⓔ
2. Ⓐ Ⓑ Ⓒ Ⓓ Ⓔ 7. Ⓐ Ⓑ Ⓒ Ⓓ Ⓔ
3. Ⓐ Ⓑ Ⓒ Ⓓ Ⓔ 8. Ⓐ Ⓑ Ⓒ Ⓓ Ⓔ
4. Ⓐ Ⓑ Ⓒ Ⓓ Ⓔ 9. Ⓐ Ⓑ Ⓒ Ⓓ Ⓔ
5. Ⓐ Ⓑ Ⓒ Ⓓ Ⓔ 10. Ⓐ Ⓑ Ⓒ Ⓓ Ⓔ

1. Y varies inversely with X. When Y is 12, X is 3. What is X when Y is 6?

   (A) 3
   (B) 6
   (C) 9
   (D) 12
   (E) 18

2. Y is inversely proportional to X. When Y is 18, X is 3. What is Y when X is 9?

   (A) 54
   (B) 36
   (C) 18
   (D) 12
   (E) 6

3. Y varies inversely in relation to the square root of X. When Y is 16, X is 9. What is X when Y is 8?

   (A) $\sqrt{6}$
   (B) 6
   (C) 36
   (D) 64
   (E) 72

4. A bowl of candy is left out on Halloween for trick-or-treaters. If each trick-or-treater takes 5 pieces of candy, there is enough for 15 trick-or-treaters. How many trick-or-treaters can get candy if each kid takes only 3 pieces of candy?

   (A) 9
   (B) 12
   (C) 15
   (D) 25
   (E) 27

5. Y + 10 varies inversely with X - 1. When Y is 8, X is 3. What is Y when X is 10?

   (A) -6
   (B) -4
   (C) 0
   (D) 18
   (E) 81

3.1.3 Identifying Variation

How can you determine whether to use Direct Variation (Y/X = K) or Inverse Variation (YX = K)? Think about whether your variable values are changing in the SAME direction (Direct Variation, also known as a ratio) or if they are going in OPPOSITE directions (Inverse Variation, which is not a ratio).

SAT Example and Technique Application:

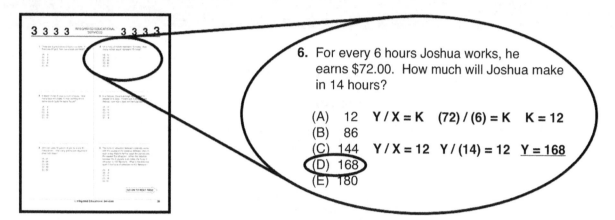

6. For every 6 hours Joshua works, he earns $72.00. How much will Joshua make in 14 hours?

   (A)   12   Y / X = K (72) / (6) = K K = 12
   (B)   86
   (C) 144   Y / X = 12 Y / (14) = 12 <u>Y = 168</u>
   (D) 168
   (E) 180

3.1.3 Let's Practice:

1. Ⓐ Ⓑ Ⓒ Ⓓ Ⓔ   6. Ⓐ Ⓑ Ⓒ Ⓓ Ⓔ
2. Ⓐ Ⓑ Ⓒ Ⓓ Ⓔ   7. Ⓐ Ⓑ Ⓒ Ⓓ Ⓔ
3. Ⓐ Ⓑ Ⓒ Ⓓ Ⓔ   8. Ⓐ Ⓑ Ⓒ Ⓓ Ⓔ
4. Ⓐ Ⓑ Ⓒ Ⓓ Ⓔ   9. Ⓐ Ⓑ Ⓒ Ⓓ Ⓔ
5. Ⓐ Ⓑ Ⓒ Ⓓ Ⓔ   10. Ⓐ Ⓑ Ⓒ Ⓓ Ⓔ

1. There are 3 girls to every 2 boys in a class. If there are 24 girls, how many boys are there?

   (A)   4
   (B)   8
   (C) 12
   (D) 16
   (E) 20

2. It takes 12 men 5 days to build a house. How many days will it take 10 men working at the same rate to build the same house?

   (A)   4
   (B)   6
   (C)   8
   (D) 10
   (E) 12

3. Jim's car uses 16 gallons of gas for every 80 miles driven. How many gallons are required to drive 140 miles?

   (A) 20
   (B) 24
   (C) 28
   (D) 30
   (E) 32

4. On a map, 2 inches represent 15 miles. How many inches would represent 75 miles?

   (A) 10
   (B) 15
   (C) 20
   (D) 25
   (E) 30

5. In a lifeboat, there is enough food to feed 4 people for 6 days. If there are 3 people in the lifeboat, how many days will the food last?

   (A) 15
   (B) 14
   (C) 12
   (D) 10
   (E)   8

6. The force of attraction between 2 planets varies with the square of the distance between them in such a way that the farther apart the planets are, the weaker the attraction. When the distance between the 2 planets is 20 miles, the force of attraction is 100 Newtons. What is the distance apart if the force of attraction is 400 Newtons?

   (A) -6
   (B) -4
   (C) 10
   (D) 18
   (E) 81

GO ON TO NEXT PAGE →

3.2 Combining Ratios

When dealing with multiple ratios that share a common variable, manipulate the ratios to make the common variable identical in both. This will allow you to compare the other parts of the ratio.

Demonstration Example

Demo: The ratio A:B is 2:7 and the ratio B:C is 4:1. Find the ratio A:C?

$\underline{A : B : C}$
2 : 7   *...put the first ratio into this row.*
    4 : 1   *...put the second ratio into this row.*

B is common to both ratios, so make it the same in both rows:
4x7 = 28, so...

$\underline{A : B : C}$
8 : 28   *...multiply the first row by 4 to get 8:28.*
    28 : 7   *...multiply the second row by 7 to get 28:7.*

8 : 28 : 7   *...Once the Bs are the same, you can conjoin the ratios. A:C = $\underline{8:7}$!*

Or, you can try writing the ratios as quotients: $\quad \dfrac{A}{B} = \dfrac{2}{7} \quad$ *and* $\quad \dfrac{B}{C} = \dfrac{4}{1}$

Notice, if you multiply across, the Bs cancel and you are left with A/C.

So, Numerator, 2 x 4 = 8 , Denominator, 7 x 1 = 7 , So, the answer is $\underline{8/7}$!

SAT Example and Technique Application:

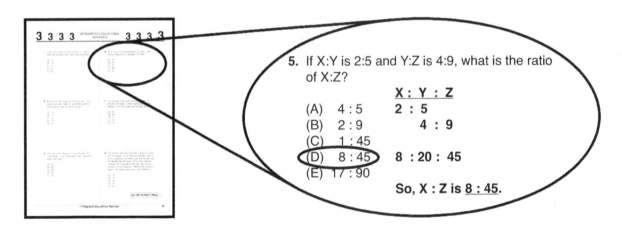

5. If X:Y is 2:5 and Y:Z is 4:9, what is the ratio of X:Z?

$\underline{X : Y : Z}$

(A) 4 : 5   2 : 5
(B) 2 : 9   4 : 9
(C) 1 : 45
(D) 8 : 45   8 : 20 : 45
(E) 17 : 90

So, X : Z is $\underline{8 : 45}$.

3.2 **Let's Practice:**

1. Ⓐ Ⓑ Ⓒ Ⓓ Ⓔ   6. Ⓐ Ⓑ Ⓒ Ⓓ Ⓔ
2. Ⓐ Ⓑ Ⓒ Ⓓ Ⓔ   7. Ⓐ Ⓑ Ⓒ Ⓓ Ⓔ
3. Ⓐ Ⓑ Ⓒ Ⓓ Ⓔ   8. Ⓐ Ⓑ Ⓒ Ⓓ Ⓔ
4. Ⓐ Ⓑ Ⓒ Ⓓ Ⓔ   9. Ⓐ Ⓑ Ⓒ Ⓓ Ⓔ
5. Ⓐ Ⓑ Ⓒ Ⓓ Ⓔ   10. Ⓐ Ⓑ Ⓒ Ⓓ Ⓔ

1. M:P is 7:8 and P:R is 2:5. What is M:R?

(A) 5 : 7
(B) 7 : 5
(C) 7 : 20
(D) 7 : 40
(E) 14 : 5

2. $3A = 4B$ and $5A = 12C$. What is B:C?

(A) 4 : 9
(B) 5 : 9
(C) 9 : 5
(D) 9 : 4
(E) 27 : 12

3. A:B is 3:5. B:C is 10:12. What is A:C?

(A) 1 : 2
(B) 1 : 3
(C) 1 : 4
(D) 2 : 1
(E) 3 : 1

4. $a/b = 4/5$ and $b/c = 1/2$. What is the value of a/c?

(A) 5 / 1
(B) 4 / 1
(C) 2 / 1
(D) 2 / 5
(E) 4 / 5

5. The ratio of x to y is 3 to 5. The ratio of x to z is 4 to 10. What is the ratio of x to y+z?

(A) 3 : 10
(B) 6 : 25
(C) 8 : 25
(D) 9 : 25
(E) 12: 25

6. M:P is 4:5. M:L is 3:4. What is M:(P+L)?

(A) 12 : 31
(B) 12 : 11
(C) 15 : 16
(D) 1 : 21
(E) 5 : 21

7. $x/y = 2/5$ and $z/y = 1/2$. What is the value of x/z?

(A) 4 / 5
(B) 5 / 4
(C) 2 / 5
(D) 5 / 2
(E) 1 / 5

GO ON TO NEXT PAGE ▷

3.3 Parts-to-the-Whole

If the ratio of x to y is 5:9 and the sum of x and y together is equal to T, you can set up a ratio as follows:

x : y

(T) $\frac{5}{14}$: $\frac{9}{14}$ (T)

SAT Example and Technique Application:

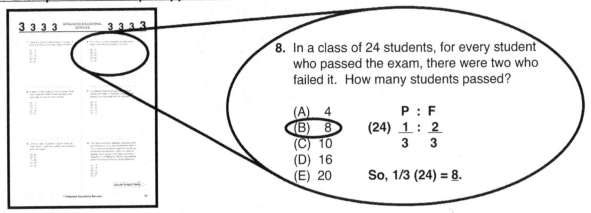

8. In a class of 24 students, for every student who passed the exam, there were two who failed it. How many students passed?

(A) 4
(B) 8
(C) 10
(D) 16
(E) 20

     P : F
(24) $\frac{1}{3}$: $\frac{2}{3}$

So, 1/3 (24) = 8.

3.3 Let's Practice:

1. Ⓐ Ⓑ Ⓒ Ⓓ Ⓔ   6. Ⓐ Ⓑ Ⓒ Ⓓ Ⓔ
2. Ⓐ Ⓑ Ⓒ Ⓓ Ⓔ   7. Ⓐ Ⓑ Ⓒ Ⓓ Ⓔ
3. Ⓐ Ⓑ Ⓒ Ⓓ Ⓔ   8. Ⓐ Ⓑ Ⓒ Ⓓ Ⓔ
4. Ⓐ Ⓑ Ⓒ Ⓓ Ⓔ   9. Ⓐ Ⓑ Ⓒ Ⓓ Ⓔ
5. Ⓐ Ⓑ Ⓒ Ⓓ Ⓔ   10. Ⓐ Ⓑ Ⓒ Ⓓ Ⓔ

1. The ratio of boys to girls is 5:6. How many girls are there if there are 220 boys and girls total?

(A) 80
(B) 100
(C) 120
(D) 140
(E) 160

2. In a pride, there is one male lion for every four lionesses. How many males are in a pride of 25?

(A) 5
(B) 8
(C) 10
(D) 16
(E) 20

3. A rectangle with a perimeter of 60 feet has a width-to-length ratio of 2:3. What is the area of the rectangle?

(A) 144
(B) 180
(C) 192
(D) 208
(E) 216

4. Over five consecutive days, a restaurant sells twice as many pizzas each day as it does on the previous day. If a total of 124 pizzas are sold over the five days, how many pizzas are sold on the last day?

(A) 16
(B) 24
(C) 32
(D) 64
(E) 124

GO ON TO NEXT PAGE ▷

3.4 Maintaining or Changing Ratios

Demonstration Examples

Maintaining...

Demo 1: On a pearl necklace, there is a 1:6 ratio of white pearls to black pearls. There are 42 pearls currently on the necklace. If 6 black pearls are added, how many white pearls must be added to maintain the ratio?

If you use Parts-to-the-whole, you will find that there are currently 6 white pearls and 36 black pearls on the necklace. 6 black pearls are added to make 42 black pearls in total. We need to add "x" white pearls to maintain the 1:6 ratio:

$$\frac{W}{B} = \frac{(6+x)}{36+6} = \frac{(6+x)}{42} = \frac{1}{4} \qquad \textbf{So, } \underline{x = 1 \text{ white pearl.}}$$

Changing...

Demo 2: How many white pearls must be added to the 42 original pearls in order to create a 1:4 ratio?

$$\frac{W}{B} = \frac{(6+x)}{36} = \frac{1}{4} \qquad \textbf{So, } \underline{x = 3 \text{ white pearls.}}$$

SAT Example and Technique Application:

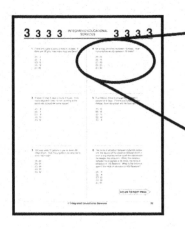

7. The ratio of tulips to violets is 5:6 and the total number of flowers is 99. If 10 tulips are added, how many violets must be added to maintain the current ratio?

(A) 12
(B) 11
(C) 10
(D) 9
(E) 8

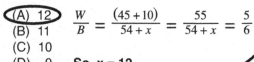

$$\frac{W}{B} = \frac{(45+10)}{54+x} = \frac{55}{54+x} = \frac{5}{6}$$

So, $\underline{x = 12.}$

3.4 Let's Practice:

1. Ⓐ Ⓑ Ⓒ Ⓓ Ⓔ   6. Ⓐ Ⓑ Ⓒ Ⓓ Ⓔ
2. Ⓐ Ⓑ Ⓒ Ⓓ Ⓔ   7. Ⓐ Ⓑ Ⓒ Ⓓ Ⓔ
3. Ⓐ Ⓑ Ⓒ Ⓓ Ⓔ   8. Ⓐ Ⓑ Ⓒ Ⓓ Ⓔ
4. Ⓐ Ⓑ Ⓒ Ⓓ Ⓔ   9. Ⓐ Ⓑ Ⓒ Ⓓ Ⓔ
5. Ⓐ Ⓑ Ⓒ Ⓓ Ⓔ   10. Ⓐ Ⓑ Ⓒ Ⓓ Ⓔ

1. In a battalion, there are 5 cavalry to every 12 infantry. When 25 cavalry are added, how many infantry must be added to the battalion to maintain the ratio?

(A) 24
(B) 36
(C) 48
(D) 60
(E) 72

2. A ratio a:b must be maintained at all times. If k is added to b, what must be added to a?

(A) k
(B) ak
(C) ak/b
(D) ab/k
(E) b/ak

3. A nuclear scientist is creating a new element. He starts with an element that has a proton-to-neutron ratio that is 13 to 71. He then sets off a nuclear reaction in which some neutrons are converted to energy and lost. The new proton-to-neutron ratio is 1 to 5. How many neutrons were lost?

(A) 70
(B) 65
(C) 13
(D) 8
(E) 6

4. A fruit punch is made from passionfruit and cranberry juices mixed together. For every 2 parts passionfruit, there are 3 parts cranberry juice. How much passionfruit juice should be added to a 25 ounce mixture in order to make the ratio 2:1?

(A) 25
(B) 20
(C) 15
(D) 10
(E) 5

3.5 Combined Averages

Demonstration Example

Demo: Class A averaged 75 on a test and class B averaged 90 on the same test. The combined average is 80. If the total number of students in both classes is 30, how many students are in class A?

A	B
75 - - - 80 - - - - - - - - 90	

Draw a number line with both averages at the ends. Place the combined average on the line.

   (5)   (10)

Take note of the different intervals.

A : B   A : B
10 : 5   2 : 1

Since the combined average is closer to 75, 75 is the larger class. So, cross the interval values to create the ratio.

$$\frac{2}{3}(30) = 20$$

Then use Parts-to-the-Whole on A's ratio to find the answer, which is <u>20 students</u>.

****<u>Note</u>: You may need to divide to get the average.**

SAT Example and Technique Application:

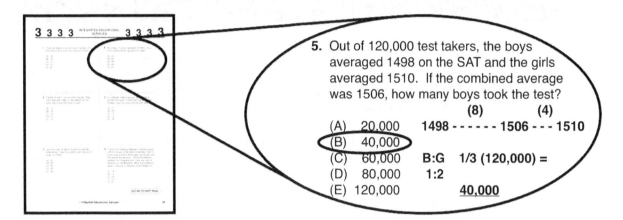

5. Out of 120,000 test takers, the boys averaged 1498 on the SAT and the girls averaged 1510. If the combined average was 1506, how many boys took the test?

                  (8)   **(4)**

(A) 20,000   1498 - - - - - - 1506 - - - 1510
(B) 40,000
(C) 60,000   B:G   1/3 (120,000) =
(D) 80,000   1:2
(E) 120,000   **40,000**

3.5 Let's Practice:

1. Ⓐ Ⓑ Ⓒ Ⓓ Ⓔ   6. Ⓐ Ⓑ Ⓒ Ⓓ Ⓔ
2. Ⓐ Ⓑ Ⓒ Ⓓ Ⓔ   7. Ⓐ Ⓑ Ⓒ Ⓓ Ⓔ
3. Ⓐ Ⓑ Ⓒ Ⓓ Ⓔ   8. Ⓐ Ⓑ Ⓒ Ⓓ Ⓔ
4. Ⓐ Ⓑ Ⓒ Ⓓ Ⓔ   9. Ⓐ Ⓑ Ⓒ Ⓓ Ⓔ
5. Ⓐ Ⓑ Ⓒ Ⓓ Ⓔ   10. Ⓐ Ⓑ Ⓒ Ⓓ Ⓔ

1. At Edison High School the juniors averaged 1850 on the SAT whereas the seniors averaged 2270. If 140 juniors and seniors took the test and the combined average was 2150, how many of these 140 students were seniors?

(A) 20
(B) 40
(C) 60
(D) 100
(E) 120

2. In a class, the boys averaged 50% on a test while the girls averaged 90% on the same test. If the combined average was 80%, then boys make up what fraction of the class?

(A) 1/3
(B) 1/4
(C) 2/3
(D) 3/4
(E) Not enough information

3. TWA sold 400 tickets for a total of $32,000. The price of a first-class ticket was $90 and the price of an economy-class ticket was $40. How many first-class tickets were sold?

(A) 320
(B) 240
(C) 160
(D)   80
(E)   40

4. Hana purchased 60 tulips and roses for a total of $600. The roses cost $12 each and the tulips cost $7 each. How many tulips did Hana buy?

(A) 10
(B) 12
(C) 16
(D) 18
(E) 24

5. Tickets to a concert are sold in advance for $5, or on the day of the concert for $8. If the combined average for all ticket sales was $6 per ticket and 100 tickets were sold on the day of the concert, how many tickets were sold in advance?

(A) 120
(B) 145
(C) 150
(D) 180
(E) 200

GO ON TO NEXT PAGE ▷

3.6 Price Ratio

Demonstration Examples

Demo 1: Jim spent $460 on two types of paint: Deluxe, priced at $3 a pint, and Standard, priced at $2 a pint. The ratio of Deluxe to Standard for Jim's purchase was 5 pints to 4 pints. In total, how many pints of paint did Jim buy?

First, enter the information into a table as shown.

	Deluxe	Standard	TOTAL:
Price/Pint	3	2	
Ratio (# of Pints)	5x	4x	
TOTAL:			460

Next, fill in the empty boxes by multiplying or adding.

	Deluxe	Standard	TOTAL:
Price/Pint	3	2	----
Ratio (# of Pints)	5x	4x	9x
TOTAL:	15x	8x	460

Once you find x, you can plug it back in to find all the other quantities.

23x = 460, x = 20, Total pints is 9x. So, 9(20) = <u>180 pints in total</u>.

Demo 2: Tickets were sold for a movie on both regular, sized screens and IMAX screens. 3 times as many tickets were sold for the regular screens as were sold for the IMAX screens. IMAX tickets cost 2 times as much as regular tickets. The total earnings from sales of both kinds of tickets on a particular day at a particular theater was $6000. How much money was brought in by IMAX tickets alone?

This time, both price and ratio require variables...

	Regular	IMAX	TOTAL:
Price/Ticket	x	2x	----
Ratio (# of Tix)	3y	y	4y
TOTAL:	3xy	2xy	6000

Once you find xy, you can plug it in to get the answer.

3xy + 2xy = 6000, xy = 1200. IMAX earnings = 2xy. So, 2(1200) = <u>$2400</u>.

3.6 **Let's Practice:**

1. Ⓐ Ⓑ Ⓒ Ⓓ Ⓔ   6. Ⓐ Ⓑ Ⓒ Ⓓ Ⓔ
2. Ⓐ Ⓑ Ⓒ Ⓓ Ⓔ   7. Ⓐ Ⓑ Ⓒ Ⓓ Ⓔ
3. Ⓐ Ⓑ Ⓒ Ⓓ Ⓔ   8. Ⓐ Ⓑ Ⓒ Ⓓ Ⓔ
4. Ⓐ Ⓑ Ⓒ Ⓓ Ⓔ   9. Ⓐ Ⓑ Ⓒ Ⓓ Ⓔ
5. Ⓐ Ⓑ Ⓒ Ⓓ Ⓔ   10. Ⓐ Ⓑ Ⓒ Ⓓ Ⓔ

1. A theater sold $300 worth of tickets on a single day. The floor tickets cost $4 and the balcony tickets cost $9. The ratio of floor to balcony tickets was 3 to 2. How many balcony tickets were sold?

(A) 20
(B) 30
(C) 40
(D) 50
(E) 60

2. Three types of gas are sold: Premium at $3.50 per gallon, Super at $3.00 per gallon, and Regular at $2.00 per gallon. If the ratio for amounts sold is 1:2:3, respectively, and $620.00 is earned, how many gallons of Super were sold?

(A) 240
(B) 120
(C) 80
(D) 60
(E) 40

3. Three types of floral arrangements were sold on Valentine's Day. The Empress bouquet has 21 roses, the Queen has 12, and the Princess has 9. A total of 990 roses were sold. Three times as many Princess bouquets as Queen bouquets were sold and twice as many Queen bouquets as Empress bouquets were sold. How many roses were sold in the Queen bouquets only?

(A) 90
(B) 120
(C) 180
(D) 240
(E) 420

4. Coke costs 3 times as much as Sprite per can. Pete's Club sold 4 times as many cans of Sprite as cans of Coke. If the total sales were $84, what was the dollar amount of the Coke sales?

(A) 24
(B) 36
(C) 48
(D) 60
(E) 72

5. Tickets to *A Midsummer Night's Dream* and *The Tempest* were being sold for $8 and $10 respectively. The ratio of tickets sold was 3:4, and the total sales reached $704. How many tickets were sold for *A Midsummer Night's Dream*?

(A) 11
(B) 22
(C) 33
(D) 44
(E) 55

6. In a trivia game, a player earns a colored game token for every question he or she answers correctly. Red, green, and blue tokens are worth 5, 10, and 15 points, respectively, and reflect the difficulty of the questions answered. A certain player has 3 times as many red tokens as green tokens and twice as many green tokens as blue tokens. If this player has scored 195 points in total, how many red game tokens does the player have?

(A) 1
(B) 3
(C) 6
(D) 9
(E) 18

GO ON TO NEXT PAGE ▷

3.7 Minimum / Maximum

Demonstration Example

Demo: If x and y are integers, x + y < 1000, and x/y = 0.375, what is the maximum value for y?

This looks a lot like Parts-to-the-Whole, except that you are not given an exact value for the total. You are only told that the total is less than 1000. Regardless, start by treating this like a Parts-to-the-Whole problem.

0.375 --> 3/8, the ratio of x:y is 3:8. So, the total number will be 11. You can then divide 1000 by 11. Since the sum must stay below 1000, you must round down to the next whole number, even if you get an integer answer. Finally, multiply by the y part of the ratio, 8.

1000 ÷ 11 = 90.9 So, we round down to 90. Then, 90(8) = 720.

SAT Example and Technique Application:

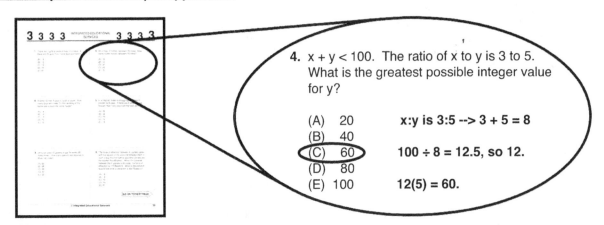

4. x + y < 100. The ratio of x to y is 3 to 5. What is the greatest possible integer value for y?

(A) 20
(B) 40
(C) 60
(D) 80
(E) 100

x:y is 3:5 --> 3 + 5 = 8

100 ÷ 8 = 12.5, so 12.

12(5) = 60.

3.7 Let's Practice:

1. Ⓐ Ⓑ Ⓒ Ⓓ Ⓔ   6. Ⓐ Ⓑ Ⓒ Ⓓ Ⓔ
2. Ⓐ Ⓑ Ⓒ Ⓓ Ⓔ   7. Ⓐ Ⓑ Ⓒ Ⓓ Ⓔ
3. Ⓐ Ⓑ Ⓒ Ⓓ Ⓔ   8. Ⓐ Ⓑ Ⓒ Ⓓ Ⓔ
4. Ⓐ Ⓑ Ⓒ Ⓓ Ⓔ   9. Ⓐ Ⓑ Ⓒ Ⓓ Ⓔ
5. Ⓐ Ⓑ Ⓒ Ⓓ Ⓔ   10. Ⓐ Ⓑ Ⓒ Ⓓ Ⓔ

1. a + b < 500 and a/b = 0.125. What is the greatest possible integer value of a?

(A) 55
(B) 110
(C) 125
(D) 390
(E) 440

2. 7a = 5b and a + b > 1000. What is the least possible integer value of a?

(A) 415
(B) 420
(C) 581
(D) 588
(E) 640

GO ON TO NEXT PAGE >

3.8 Sequences

3.8.1 **Arithmetic Sequences**

To find a term in an arithmetic sequence, use this equation: $a_n = a_1 + d(n - 1)$

a_n = the nth term a_1 = the 1st term   **d** = the difference between terms   **n** = the term's number

To find the sum of the first "n" terms of an arithmetic sequence, use this equation: $s_n = n(a_1 + a_n) / 2$

SAT Example and Technique Application:

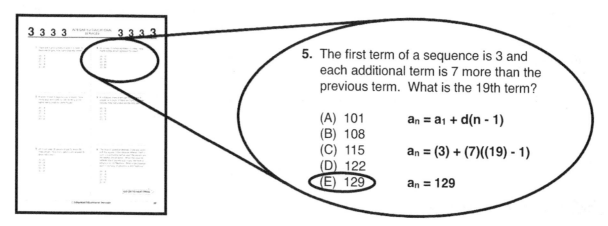

5. The first term of a sequence is 3 and each additional term is 7 more than the previous term. What is the 19th term?

(A) 101 $a_n = a_1 + d(n - 1)$
(B) 108
(C) 115 $a_n = (3) + (7)((19) - 1)$
(D) 122
(E) 129 $a_n = 129$

3.8.1 **Let's Practice:**

1. (A)(B)(C)(D)(E)   6. (A)(B)(C)(D)(E)
2. (A)(B)(C)(D)(E)   7. (A)(B)(C)(D)(E)
3. (A)(B)(C)(D)(E)   8. (A)(B)(C)(D)(E)
4. (A)(B)(C)(D)(E)   9. (A)(B)(C)(D)(E)
5. (A)(B)(C)(D)(E)   10. (A)(B)(C)(D)(E)

1. The first term of a sequence is -9 and each term increases by 3. What is the 99th term?

(A) 279
(B) 282
(C) 285
(D) 288
(E) 291

2. In the sequence -28, -12, 4, 20 ..., which term has a value of 388?

(A) 24th
(B) 25th
(C) 26th
(D) 27th
(E) 28th

3. The first term of a sequence is 5 and each term after the first term is 3 more than the preceding term. What is the sum of the first 21 terms of this sequence?

(A) 735
(B) 738
(C) 741
(D) 744
(E) 747

4. What is the sum of the first 100 positive integers?

(A) 5000
(B) 5050
(C) 5100
(D) 5200
(E) 5500

GO ON TO NEXT PAGE ⟩

3 **3** **3** **3**

Unauthorized copying or
reuse of any part of this
page is illegal.

3 **3** **3** **3**

5. What is the sum of the first hundred positive odd integers?

(A) 5,000
(B) 5,050
(C) 10,000
(D) 10,100
(E) 10,500

6. A is the sum of all positive even integers up to and including 88. B is the sum of all positive odd integers up to and including 89. What is A + B?

(A) 3995
(B) 4000
(C) 4005
(D) 4048
(E) 4089

7. X is the sum of all positive integers from 7 to 77. Y is the sum of all positive integers from 6 to 76. What is X - Y?

(A) 60
(B) 61
(C) 70
(D) 71
(E) 83

8. G is the sequence made up of all positive even integers. H is the sequence of consecutive positive integers starting at 1. When each term in G is added to each term in H, the result is a new sequence called K. What is the 17th term in the sequence K?

(A) 50
(B) 51
(C) 60
(D) 61
(E) 70

3.8.2 **Geometric Sequences**

To find a term in a geometric sequence, use this equation: $a_n = (a_1)r^{(n-1)}$

a_n = the nth term a_1 = the 1st term r = the ratio of one term to the next n = the term's number

SAT Example and Technique Application:

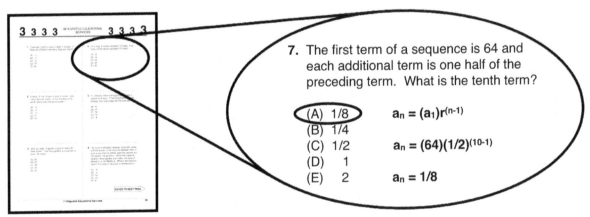

7. The first term of a sequence is 64 and each additional term is one half of the preceding term. What is the tenth term?

(A) 1/8 $a_n = (a_1)r^{(n-1)}$
(B) 1/4
(C) 1/2 $a_n = (64)(1/2)^{(10-1)}$
(D) 1
(E) 2 $a_n = 1/8$

3.8.2 **Let's Practice:**

1. Ⓐ Ⓑ Ⓒ Ⓓ Ⓔ   6. Ⓐ Ⓑ Ⓒ Ⓓ Ⓔ
2. Ⓐ Ⓑ Ⓒ Ⓓ Ⓔ   7. Ⓐ Ⓑ Ⓒ Ⓓ Ⓔ
3. Ⓐ Ⓑ Ⓒ Ⓓ Ⓔ   8. Ⓐ Ⓑ Ⓒ Ⓓ Ⓔ
4. Ⓐ Ⓑ Ⓒ Ⓓ Ⓔ   9. Ⓐ Ⓑ Ⓒ Ⓓ Ⓔ
5. Ⓐ Ⓑ Ⓒ Ⓓ Ⓔ   10. Ⓐ Ⓑ Ⓒ Ⓓ Ⓔ

1. The first term in a sequence is 3 and each term after the first term is twice the preceding term. What is the 7th term?

    (A)   48
    (B)   96
    (C) 192
    (D) 384
    (E) 768

2. In a sequence, each term is 1.5 times the preceding term. If the first term is 8, which term in the sequence has a value of 27?

    (A) 4th
    (B) 5th
    (C) 6th
    (D) 7th
    (E) 8th

3. The first term in a sequence is 7 and each term that follows is -2 times the preceding term. What is the sum of the first 5 terms?

    (A) 70
    (B) 77
    (C) 84
    (D) 91
    (E) 98

4. What is the first term in a series in which each term is 4 times the preceding term and in which the 7th term is 2048?

    (A) 1/8
    (B) 1/2
    (C) 2
    (D) 8
    (E) 32

5. In a sequence that begins with 1 and in which each term is -1 times the preceding term, what is the sum of the first 100 terms?

    (A)   2
    (B)   1
    (C)   0
    (D) -1
    (E) -2

6. In a sequence that begins with 5 and in which each term is -1 times the preceding term, what is the sum of the first 99 terms?

    (A) 10
    (B)   5
    (C)   0
    (D) -5
    (E) -10

7. In the sequence 5, 15, 45, ..., each term is 3 times the term before it. Which of the following is an expression for the 100th term in this sequence?

    (A) $5(100)$
    (B) $5(5)^{100}$
    (C) $5(3)^{100}$
    (D) $5(3)^{99}$
    (E) $3(5)^{99}$

8. In the sequence 4, 8, 16, ..., each term is twice the preceding term. Which of the following is an expression for the nth term in the sequence?

    (A) $2(n + 1)$
    (B) $2n$
    (C) 2^n
    (D) $2^{n + 1}$
    (E) $2^{n - 1}$

9. Given $A_n = 4(7)^{9 - 1}$, where 4 is the first term, which term in the geometric sequence is defined by A_n?

    (A) 4th term
    (B) 5th term
    (C) 7th term
    (D) 8th term
    (E) 9th term

GO ON TO NEXT PAGE ⇒

3.8.3 Growth and Decay

For a substance that doubles in mass every 6 hours, the GROWTH model is: $M_t = M_0(2)^{t/6}$

M_t = Mass at time t M_0 = Mass at time 0 t = time

For a substance that decreases in mass by 36% every 3 hours, the DECAY model is: $M_t = M_0(0.64)^{t/3}$

M_t = Mass at time t M_0 = Mass at time 0 t = time

SAT Example and Technique Application:

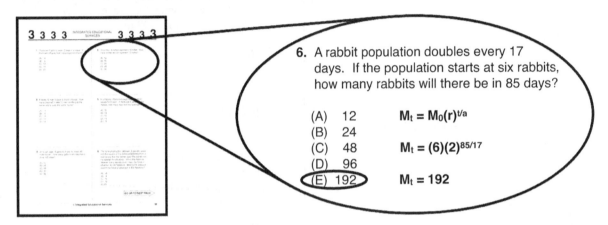

6. A rabbit population doubles every 17 days. If the population starts at six rabbits, how many rabbits will there be in 85 days?

   (A)   12 $M_t = M_0(r)^{t/a}$
   (B)   24
   (C)   48 $M_t = (6)(2)^{85/17}$
   (D)   96
   (E)   192 $M_t = 192$

3.8.3 Let's Practice:

1. (A)(B)(C)(D)(E)   6. (A)(B)(C)(D)(E)
2. (A)(B)(C)(D)(E)   7. (A)(B)(C)(D)(E)
3. (A)(B)(C)(D)(E)   8. (A)(B)(C)(D)(E)
4. (A)(B)(C)(D)(E)   9. (A)(B)(C)(D)(E)
5. (A)(B)(C)(D)(E)   10. (A)(B)(C)(D)(E)

1. The price of bread increases by 5% every 6 years. If the price is $2 in Year 1, what is the approximate price in Year 30?

   (A) $2.40
   (B) $2.45
   (C) $2.50
   (D) $2.55
   (E) $2.60

2. A log decays at the rate of 20% every 5 years. If a 500 kg log is left to decay over 20 years, what will its mass be at the end of this period?

   (A) 204.8
   (B) 204.1
   (C) 102.8
   (D) 102.4
   (E) 100.0

3. A population of mosquitos quadruples every 60 days. If there are 8 mosquitos today, how many mosquitos will there be in 300 days?

   (A)   128
   (B)   512
   (C)   2048
   (D)   8192
   (E)   32768

GO ON TO NEXT PAGE >

4. A culture of bacteria is grown in a laboratory in a petri dish. The culture triples in mass every 4 days. After 16 days, the mass of the new culture is what percent of the mass of the old culture?

(A)   9%
(B)   81%
(C)   810%
(D) 8100%
(E) 9000%

5. If a sum of $1000 is placed in an account where the account will grow 6% annually, how much will the account have grown to after 10 years have passed?

(A) $79.08
(B) $790.80
(C) $1,079.08
(D) $1,790.85
(E) $1,801.80

6. A radioactive isotope has a half-life of 50 years. its mass in 1980 was 800 grams. What will its mass be in the year 2020?

(A) 459.5 g
(B) 455.9 g
(C) 449.5 g
(D) 449.4 g
(E) 409.5 g

7. In a forensic lab, there is a machine that can double a sample of tissue every half hour. If there is 1 gram of tissue at the beginning, how much tissue is there after 4 hours?

(A)   64 g
(B)   128 g
(C)   256 g
(D)   512 g
(E) 1024 g

3.8.4 **Other Sequences**

Sequence problems on the SAT will not always fall neatly into the categories of Arithmetic or Geometric Sequences. The SAT can always define a new type of sequence and expect you to understand and answer a relevant question. Just write out the sequence if you are only dealing with a few terms, or look for a shortcut.

Demonstration Example

Demo: The first term in a sequence of integers is 2. Every term after that is the product of the preceding term and -3. What is the first term of the sequence that is greater than 100?

This is a geometric sequence, but it is not clear how a formula could help us here. Just generate the terms of the sequence as indicated in the question:

2, -6, 18, -54, 162, ... So, the 5th term is the first term that is over 100.

SAT Example and Technique Application:

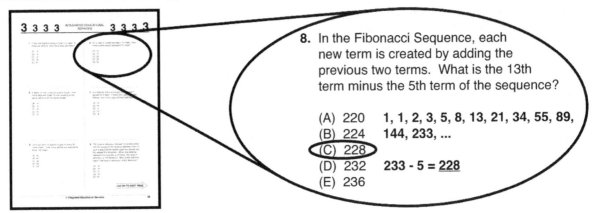

8. In the Fibonacci Sequence, each new term is created by adding the previous two terms. What is the 13th term minus the 5th term of the sequence?

(A) 220
(B) 224
(C) 228
(D) 232
(E) 236

1, 1, 2, 3, 5, 8, 13, 21, 34, 55, 89, 144, 233, ...

233 - 5 = **228**

3.8.4 **Let's Practice:**

1. (A)(B)(C)(D)(E)   6. (A)(B)(C)(D)(E)
2. (A)(B)(C)(D)(E)   7. (A)(B)(C)(D)(E)
3. (A)(B)(C)(D)(E)   8. (A)(B)(C)(D)(E)
4. (A)(B)(C)(D)(E)   9. (A)(B)(C)(D)(E)
5. (A)(B)(C)(D)(E)   10. (A)(B)(C)(D)(E)

1. The first two terms of a sequence are 3 and 4. Every term that follows is generated by taking the non-negative difference of the previous terms. After which term is there a repeating pattern?

(A) 4
(B) 5
(C) 6
(D) 7
(E) 8

2. The nth term in a certain sequence is equal to $(n + 1)(n - 1)$. How much larger is the 16th term than the 12th term?

(A) 112
(B) 87
(C) 60
(D) 31
(E) 0

3. Let $S = \{ 1 + 2 + ... + 40 \}$ and let $T = \{ 41 + 42 + ... + 80 \}$. What is T - S?

(A) 1200
(B) 1400
(C) 1600
(D) 1800
(E) 2000

4. Sequence 1 = { 2, 4, 6, 8, ... }
   Sequence 2 = { 8, 10, 12, 14, ... }

   Both sequences are continued to their 20th terms. Let K be the sum of all terms in Sequence 1, and let M be the sum of all the terms in Sequence 2. What is M - K?

(A) 120
(B) 112
(C) 104
(D) 100
(E) 96

5. The first 5 terms of a certain sequence are: 1, 3, 7, 15, 31, ... What will the 10th term of the sequence be?

(A) 63
(B) 127
(C) 255
(D) 511
(E) 1023

6. The first 4 terms of a sequence are: 2, 3, 8, 63, ... What is the 5th term?

(A) 125
(B) 503
(C) 3968
(D) 3970
(E) 250047

GO ON TO NEXT PAGE ▷

CHAPTER 3: CHALLENGE QUESTIONS

Student-Produced Responses

3.1

1. If the square of a certain number is directly proportional to the number minus one, What is the only possible value for the number if k, the constant multiplier, is 4?

3.2

2. Suppose the following:

   The ratio of A:B is 3:5.
   The ratio of A:C is 2:7
   The ratio of C:D is 3:1.
   The ratio of D:E is 1:4.

   What is the ratio of B:E?

3.4

3. Suppose that you have an ant farm where there are a total of 899 ants and that the ratio of worker ants to harvester ants is 12:17. If 400 harvester ants suddenly died, how many more harvester ants would you need to purchase if you want the final ratio to be one worker ant for every two harvester ants?

3.5

4. A certain group of children, B, has an average test score that is twice as high as the average score of the students in group A. If the combined average of both groups is a 1700, and the ratio for the number of students in class A to the number in class B is 3:7, what is the average test score for group B?

3.6

5. Suppose that garlic bagels cost twice as much as sesame bagels, which in turn cost twice as much as plain bagels. Also, suppose that the ratio of the number of bagels sold, plain:sesame:garlic, is 3:2:1. If Deborah spends a total of $44 on bagels, what was the total amount of money Deborah spent on plain bagels?

3.8

6. The first 7 terms in a sequence are as follows...

   $$0 , 1 , 3 , 2 , 8 , 3 , 15 , ...$$

   What is the sum of the 8th and 9th terms?

GO ON TO NEXT PAGE >

| Multiple-Choice | Student-Produced Responses |

CHAPTER 3 REVIEW

1 Ⓐ Ⓑ Ⓒ Ⓓ Ⓔ
2 Ⓐ Ⓑ Ⓒ Ⓓ Ⓔ
3 Ⓐ Ⓑ Ⓒ Ⓓ Ⓔ
4 Ⓐ Ⓑ Ⓒ Ⓓ Ⓔ
Ⓐ Ⓑ Ⓒ Ⓓ Ⓔ
Ⓐ Ⓑ Ⓒ Ⓓ Ⓔ
Ⓐ Ⓑ Ⓒ Ⓓ Ⓔ
Ⓐ Ⓑ Ⓒ Ⓓ Ⓔ
Ⓐ Ⓑ Ⓒ Ⓓ Ⓔ
Ⓐ Ⓑ Ⓒ Ⓓ Ⓔ

5    6 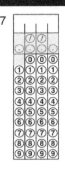   7    8    9

1. A rocket burns 200 liters of fuel to climb 1200 meters in altitude. How much fuel must be burned for the rocket to reach 3000 meters?

   (A) 300 liters
   (B) 400 liters
   (C) 500 liters
   (D) 600 liters
   (E) 700 liters

2. In a bag of coins, there are nickels, dimes and quarters. If there are 3 times as many nickels as dimes and 5 times as many dimes as quarters, and the total value of the coins in the bag is $15, how many nickels are in the bag?

   (A) 10
   (B) 20
   (C) 50
   (D) 120
   (E) 150

3. If $x/y = 3/5$ and $y/z = 7/8$, what is the value of x/z?

   (A) 7/10
   (B) 3/5
   (C) 21/40
   (D) 1/2
   (E) 1/7

4. In a toy store, the ratio of action figures to monster figurines is 4 to 5. If there are 600 action figures, how many monster figurines are there?

   (A) 700
   (B) 750
   (C) 800
   (D) 850
   (E) 1000

5. An ice cream vendor sells vanilla ice cream cones for $6 and chocolate ice cream cones for $10. If the ice cream vendor makes $700 after selling a total of 100 ice cream cones, what fraction of the sales is accounted for by vanilla ice cream cones?

6. On a stained glass window, the ratio of red to green to blue stained glass pieces is 2:4:7, respectively. If the window is made up of 182 pieces of glass, how many more blue pieces than green pieces are there?

7. A list of numbers is generated by starting with 5 and counting forward by fives. What is the value of the 26th term of this list?

8. The sum of all consecutive integers from -18 to x, inclusive, is 60. What is the value of x?

9. A sequence begins with the number 3. The ratio of each term in the sequence to the term preceding it is 4 to 1. The 12th term of this sequence has how many more digits than the 10th term of this sequence?

GO ON TO NEXT PAGE ⟩

Multiple-Choice

1 Ⓐ Ⓑ Ⓒ Ⓓ Ⓔ
2 Ⓐ Ⓑ Ⓒ Ⓓ Ⓔ
3 Ⓐ Ⓑ Ⓒ Ⓓ Ⓔ
4 Ⓐ Ⓑ Ⓒ Ⓓ Ⓔ
   Ⓐ Ⓑ Ⓒ Ⓓ Ⓔ
   Ⓐ Ⓑ Ⓒ Ⓓ Ⓔ
   Ⓐ Ⓑ Ⓒ Ⓓ Ⓔ
   Ⓐ Ⓑ Ⓒ Ⓓ Ⓔ
   Ⓐ Ⓑ Ⓒ Ⓓ Ⓔ
   Ⓐ Ⓑ Ⓒ Ⓓ Ⓔ

Student-Produced Responses

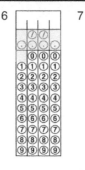

CHAPTER
1 - 3
CUMULATIVE REVIEW

1. $x + 3 = y$. What is $x^2 - 9$ in terms of y?

   (A) $y - 9$
   (B) $y - 6$
   (C) $y^2 - 6y + 9$
   (D) $y^2 - 6y + 6$
   (E) $y^2 - 6y$

2. x percent of 50 is 35. What is x percent of 70?

   (A) 45
   (B) 49
   (C) 50
   (D) 51
   (E) 55

3. The average of five consecutive numbers is 35. What is the greatest of these numbers?

   (A) 37
   (B) 39
   (C) 41
   (D) 43
   (E) 45

4. A copy machine can turn out 900 copies in an hour. How many minutes will it take the copy machine to make 1600 copies?

   (A) 100.00
   (B) 105.00
   (C) 105.33
   (D) 106.67
   (E) 108.50

5. Will is buying supplies for his office. He buys twice as many boxes of staples as post-it pads and three more post-it pads than boxes of pens. If the total number of all three kinds of items combined is 25, how many post-it pads will he buy?

6. In a recipe for chocolate cake, cake mix, eggs, and sugar are required to make the batter. By weight, the ratio of cake mix to eggs to sugar is 6:3:2, respectively. How much, by weight, will eggs account for in 66 oz. of batter?

7. The force of an object's motion is equal to its mass times its acceleration ($F = ma$). If the mass of an object is increased by 10% and its acceleration is decreased by 10%, the new force of the object's motion is what percent of the old force?

8. In a wine collector's collection, there are three times as many bottles of Merlot as bottles of Cabernet Sauvignon, and each bottle of Cabernet Sauvignon costs twice as much as each bottle of Merlot. If the collector has $6000 worth of Cabernet and Merlot bottles, how much did he spend on the Cabernet?

9. The price of a newly released video game is $100. A customer uses a 15%-off-on-any-item coupon as well as a 20% employee discount through a friend who works at the store. How much did the customer pay for the game?

STOP

Chapter 4: Number Properties

8 Sections
118 Practice Questions

4.1 Rules of Divisibility

If you want to know how many integers less than 1000 are divisible by n, then you simply divide 1000 by n and drop the decimal. For example, to find out how many integers less than 1000 are divisible by 7, you divide 1000 by 7, which equals 142.8571429, then you simply ignore the decimal. The final number is 142. Never round up! You must always round down, even if it goes in evenly!

If you see the word "and," you must divide by the least common multiple of the divisors.

If you see the word "or," you must divide both separately, then subtract the "and" portion where they overlap.

Demonstration Examples

Demo 1: How many integers less than 100 are divisible by 6?

$100 \div 6 = 16.666667$ So, <u>16</u>.

Demo 2: How many integers less than 100 are divisible by 3 <u>and</u> 5?

LCM of 3 and 5 is 15. So, $100 \div 15 = 6.666667$. So, <u>6</u>.

Demo 3: How many integers less than 100 are divisible by 3 <u>or</u> 5?

$100 \div 3 = 33.333$, $100 \div 5 = 20$, $100 \div 15 = 6.667$
So, $33 + 19 - 6 = $ <u>46</u>.

SAT Example and Technique Application:

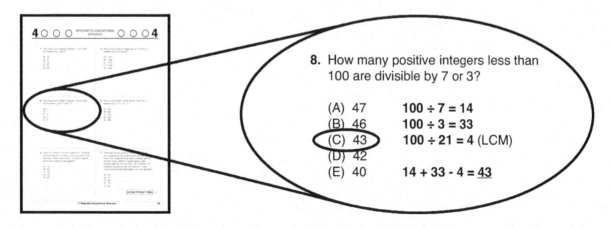

8. How many positive integers less than 100 are divisible by 7 or 3?

(A) 47 $100 \div 7 = 14$
(B) 46 $100 \div 3 = 33$
(C) 43 $100 \div 21 = 4$ (LCM)
(D) 42
(E) 40 $14 + 33 - 4 = $ <u>43</u>

4.1 **Let's Practice:**

1. Ⓐ Ⓑ Ⓒ Ⓓ Ⓔ   6. Ⓐ Ⓑ Ⓒ Ⓓ Ⓔ
2. Ⓐ Ⓑ Ⓒ Ⓓ Ⓔ   7. Ⓐ Ⓑ Ⓒ Ⓓ Ⓔ
3. Ⓐ Ⓑ Ⓒ Ⓓ Ⓔ   8. Ⓐ Ⓑ Ⓒ Ⓓ Ⓔ
4. Ⓐ Ⓑ Ⓒ Ⓓ Ⓔ   9. Ⓐ Ⓑ Ⓒ Ⓓ Ⓔ
5. Ⓐ Ⓑ Ⓒ Ⓓ Ⓔ   10. Ⓐ Ⓑ Ⓒ Ⓓ Ⓔ

1. How many even integers between 1 and 1000 are divisible by 3 and 5?

(A) 30
(B) 31
(C) 33
(D) 63
(E) 66

2. How many odd integers between 100 and 200 are divisible by both 7 and 11?

(A) 0
(B) 1
(C) 2
(D) 3
(E) 4

3. Clock A chimes on the hour every hour. Clock B chimes every 30 minutes on the hour and on the half hour. How many times in a 24-hour period do the two clocks chime together?

(A) 12
(B) 24
(C) 36
(D) 48
(E) 60

4. The sequence 3, 4, 5, 3, 4, 5, 3, 4, 5, ... is repeated. How many 3s are in the first 100 terms?

(A) 31
(B) 32
(C) 33
(D) 34
(E) 36

5. What is the smallest integer above 1000 that is divisible by both 6 and 9?

(A) 1001
(B) 1008
(C) 1018
(D) 1036
(E) 1054

6. What is the largest integer below 1000 that is divisible by 2, 3, 5, and 7?

(A) 630
(B) 800
(C) 840
(D) 950
(E) 980

7. Suitcases coming down a ramp for packaging are inspected by two workers, Alan and Bertram. While Alan inspects every fourth suitcase starting with the fourth, Bertram inspects every sixth suitcase starting with the sixth. On Tuesday, 127 suitcases passed by Alan and Bertram. How many suitcases were packaged, but not inspected?

(A) 42
(B) 47
(C) 52
(D) 75
(E) 85

GO ON TO NEXT PAGE ▷

4.2 Remainders

Demonstration Examples

Type 1: When finding the remainder R when a number N is divided by D, follow the steps defined in the examples below:

Demo 1: Find the remainder R when 45 is divided by 7.

$45 \div 7 = 6.42857$, $6.42857 - 6 = .42857$, $.42857 \times 7 = 3$. So, **R = 3**.

Demo 2: A number series is generated this: 1, 2, 3, 4, 1, 2, 3, 4, ... If the pattern continues, what will be the 50th term?

Since the pattern repeats every 4 numbers divide 50 by 4 and get 12 with a remainder of 2. This means that after twelve repetitions, the pattern will end on a 4. So, you need two more. The answer is 2.

Type 2: When finding a number N that has a remainder of R when divided by D, follow the steps defined in the example below:

Demo 3: What number N has a remainder of 5 when divided by 9?

$N = Dx + R$, $N = ?$ $D = 9$ $R = 5$ So, $N = 9x + 5$
Now, any non-negative value of x will give you an answer...
$N = 9(0) + 5 = \underline{5}$. $N = 9(1) + 5 = \underline{14}$. $N = 9(2) + 5 = \underline{23}$. Any of these will work.

SAT Example and Technique Application:

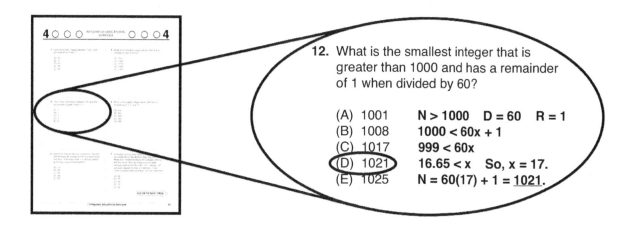

12. What is the smallest integer that is greater than 1000 and has a remainder of 1 when divided by 60?

(A) 1001
(B) 1008
(C) 1017
(D) 1021
(E) 1025

$N > 1000$ $D = 60$ $R = 1$
$1000 < 60x + 1$
$999 < 60x$
$16.65 < x$ So, $x = 17$.
$N = 60(17) + 1 = \underline{1021}$.

4.2 **Let's Practice:**

1. Ⓐ Ⓑ Ⓒ Ⓓ Ⓔ   6. Ⓐ Ⓑ Ⓒ Ⓓ Ⓔ
2. Ⓐ Ⓑ Ⓒ Ⓓ Ⓔ   7. Ⓐ Ⓑ Ⓒ Ⓓ Ⓔ
3. Ⓐ Ⓑ Ⓒ Ⓓ Ⓔ   8. Ⓐ Ⓑ Ⓒ Ⓓ Ⓔ
4. Ⓐ Ⓑ Ⓒ Ⓓ Ⓔ   9. Ⓐ Ⓑ Ⓒ Ⓓ Ⓔ
5. Ⓐ Ⓑ Ⓒ Ⓓ Ⓔ   10. Ⓐ Ⓑ Ⓒ Ⓓ Ⓔ

1. What is the remainder when 467 is divided by 12?

   (A) 10
   (B) 11
   (C) 12
   (D) 13
   (E) 14

2. What month is 535 months after May?

   (A) November
   (B) December
   (C) January
   (D) February
   (E) March

3. Find the 50th term of this sequence:
   1, 2, 3, 4, 5, 6, 7, 8, 9, 1, 2, 3, 4, 5, 6, 7, 8, 9, ...

   (A) 1
   (B) 2
   (C) 3
   (D) 4
   (E) 5

4. A card dispenser at a casino dispenses cards in the following order: Ace, King, Queen, and Jack. What will be the 73rd card dispensed by the machine?

   (A) Ace
   (B) King
   (C) Queen
   (D) Jack
   (E) The answer cannot be determined.

5. What day of the week is 64 days from Wednesday?

   (A) Thursday
   (B) Friday
   (C) Saturday
   (D) Sunday
   (E) Monday

6. What is the largest integer less than 100 that has a remainder of 5 when divided by 7?

   (A) 91
   (B) 92
   (C) 95
   (D) 96
   (E) 99

7. What is the least positive integer that has a remainder of 4 when divided by 5?

   (A) 4
   (B) 9
   (C) 14
   (D) 19
   (E) 24

8. When K is divided by 3 the remainder is 1. What is the remainder when K + 1 is divided by 3?

   (A) 0
   (B) 1
   (C) 2
   (D) 3
   (E) 4

9. What number between 0 and 30 has a remainder of 5 when divided by 9 and a remainder of 2 when divided by 7?

   (A) 9
   (B) 14
   (C) 16
   (D) 23
   (E) 30

10. What is the greatest number less than 50 that has a remainder of 3 when divided by 5 or by 7?

    (A) 36
    (B) 40
    (C) 42
    (D) 45
    (E) 48

GO ON TO NEXT PAGE ▷

4.3 Odds and Evens

ODD x ODD = ODD
Odd x Even = Even
Even x Even = Even

Odd + Odd = Even
ODD + EVEN = ODD
Even + Even = Even

SAT Example and Technique Application:

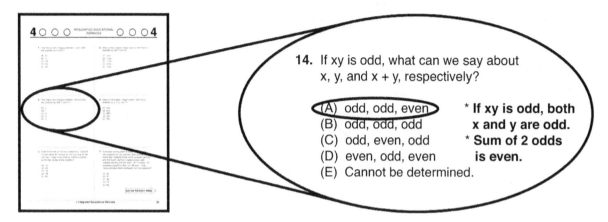

14. If xy is odd, what can we say about x, y, and x + y, respectively?

   (A) odd, odd, even
   (B) odd, odd, odd
   (C) odd, even, odd
   (D) even, odd, even
   (E) Cannot be determined.

* **If xy is odd, both x and y are odd.**
* **Sum of 2 odds is even.**

4.3 Let's Practice:

1. Ⓐ Ⓑ Ⓒ Ⓓ Ⓔ   6. Ⓐ Ⓑ Ⓒ Ⓓ Ⓔ
2. Ⓐ Ⓑ Ⓒ Ⓓ Ⓔ   7. Ⓐ Ⓑ Ⓒ Ⓓ Ⓔ
3. Ⓐ Ⓑ Ⓒ Ⓓ Ⓔ   8. Ⓐ Ⓑ Ⓒ Ⓓ Ⓔ
4. Ⓐ Ⓑ Ⓒ Ⓓ Ⓔ   9. Ⓐ Ⓑ Ⓒ Ⓓ Ⓔ
5. Ⓐ Ⓑ Ⓒ Ⓓ Ⓔ   10. Ⓐ Ⓑ Ⓒ Ⓓ Ⓔ

1. If x is the sum of 3 even integers and y is the product of 2 odd integers, which of the following is true?

   (A) x is even, y is even, their sum is even
   (B) x is even, y is odd, their sum is even
   (C) x is even, y is odd, their sum is odd
   (D) x is even, y is even, their sum is odd
   (E) Their sum cannot be classified

2. If x and y are integers and $xy + x^2$ is odd, what can we say about x, y, and their sum?

   (A) x is even, y is even, their sum is even
   (B) x is even, y is odd, their sum is odd
   (C) x is odd, y is odd, their sum is even
   (D) x is odd, y is even, their sum is odd
   (E) Their sum cannot be classified

3. If xy is even and yz is odd, what are x, y, and z?

   (A) x is even, y is odd, z is odd
   (B) x is even, y is odd, z is even
   (C) x is odd, y is even, z is even
   (D) x is odd, y is even, z is odd
   (E) x is odd, y is odd, z is odd

4. A pizza can be cut straight across into n equal slices, where n is?

   (A) Any positive integer
   (B) Any even number
   (C) Any odd number
   (D) Only multiples of 4
   (E) Only 8

5. If x is any positive integer, which of the following is an expression for an even integer that is twice the value of an odd integer?

   (A) 2x + 1
   (B) 2x + 2
   (C) 4x
   (D) 4x + 1
   (E) 4x + 2

GO ON TO NEXT PAGE ▷

4.4 Number Properties

4.4.1 Prime Numbers

Facts:
* Prime numbers can only be divided by themselves and by 1.
* 1 is NOT a prime number. 0 is NOT a prime number.
* 2 is the ONLY even prime number.
* Prime numbers can only be positive integers.
* There are 15 prime numbers less than 50. MEMORIZE THEM!

SAT Example and Technique Application:

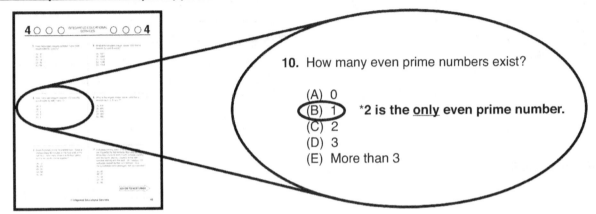

10. How many even prime numbers exist?

(A) 0
(B) 1   *2 is the **only** even prime number.
(C) 2
(D) 3
(E) More than 3

4.4.1 Let's Practice:

1. Ⓐ Ⓑ Ⓒ Ⓓ Ⓔ   6. Ⓐ Ⓑ Ⓒ Ⓓ Ⓔ
2. Ⓐ Ⓑ Ⓒ Ⓓ Ⓔ   7. Ⓐ Ⓑ Ⓒ Ⓓ Ⓔ
3. Ⓐ Ⓑ Ⓒ Ⓓ Ⓔ   8. Ⓐ Ⓑ Ⓒ Ⓓ Ⓔ
4. Ⓐ Ⓑ Ⓒ Ⓓ Ⓔ   9. Ⓐ Ⓑ Ⓒ Ⓓ Ⓔ
5. Ⓐ Ⓑ Ⓒ Ⓓ Ⓔ   10. Ⓐ Ⓑ Ⓒ Ⓓ Ⓔ

1. How many prime numbers are less than 10?

(A) 3
(B) 4
(C) 5
(D) 6
(E) 7

2. How many multiples of 7 are prime?

(A) All of them
(B) 3
(C) 2
(D) 1
(E) 0

3. A and B are prime numbers such that B = 6 + A, and no prime number exists between A and B. What is the lowest possible value of B + A?

(A) 68
(B) 65
(C) 52
(D) 40
(E) 28

---

4. A, B, and C are prime numbers such that A + B + C = 16 and 2(A + B) + 1 = C. What is the value of C?

(A) 11
(B) 13
(C) 17
(D) 19
(E) 23

GO ON TO NEXT PAGE ⟩

4.4.2 **Integers and Digits**

Facts:
* Digits can only be 0 through 9.
* Integers are made up of 1 or more digits and can be either positive or negative.
* The integer 527 is made up of a units digit of 7, a tens digit of 2, and a hundreds digit of 5.

Demonstration Example

Demo: How many positive integers less than 100 contain at least one digit 7?

1. **There are 9 x 9 = 81 possibilities that contain NO digit 7.**
2. **Subtract 1 from 81. (00 is not a positive integer.) So, 80.**
3. **Now, Subtract 80 from the 99 positive integers that are less than 100, and you get...   99 - 80 = 19.**

SAT Example and Technique Application:

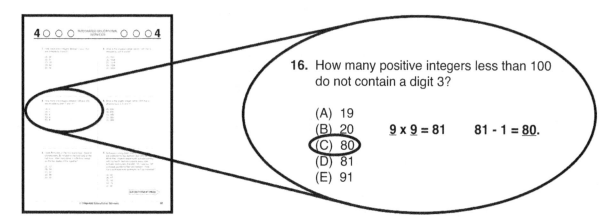

16. How many positive integers less than 100 do not contain a digit 3?

(A) 19
(B) 20
(C) 80
(D) 81
(E) 91

9 x 9 = 81   81 - 1 = 80.

4.4.2 **Let's Practice:**

1. Ⓐ Ⓑ Ⓒ Ⓓ Ⓔ   6. Ⓐ Ⓑ Ⓒ Ⓓ Ⓔ
2. Ⓐ Ⓑ Ⓒ Ⓓ Ⓔ   7. Ⓐ Ⓑ Ⓒ Ⓓ Ⓔ
3. Ⓐ Ⓑ Ⓒ Ⓓ Ⓔ   8. Ⓐ Ⓑ Ⓒ Ⓓ Ⓔ
4. Ⓐ Ⓑ Ⓒ Ⓓ Ⓔ   9. Ⓐ Ⓑ Ⓒ Ⓓ Ⓔ
5. Ⓐ Ⓑ Ⓒ Ⓓ Ⓔ   10. Ⓐ Ⓑ Ⓒ Ⓓ Ⓔ

1. How many positive integers less than 100 contain at least one digit 4?

(A) 19
(B) 20
(C) 21
(D) 80
(E) 81

2. How many positive integers less than 1000 contain no digit 5?

(A) 270
(B) 271
(C) 272
(D) 728
(E) 729

GO ON TO NEXT PAGE ⇨

3. How many positive integers less than 1000 contain at least one digit 8?

(A) 270
(B) 271
(C) 272
(D) 728
(E) 729

4. How many positive integers less than 10,000 contain at least one digit 6?

(A) 3438
(B) 3439
(C) 6559
(D) 6560
(E) 6561

5. How many positive integers less than 1,000 contain exactly 2 digit 9s?

(A) 17
(B) 19
(C) 20
(D) 21
(E) 27

6. XY is a two-digit integer that is divisible by 9. What is the maximum possible value of XY?

(A) 99
(B) 90
(C) 89
(D) 72
(E) 54

7. PR is a positive two-digit integer divisible by 5. What is the minimum possible value of PR?

(A) 0
(B) 5
(C) 10
(D) 25
(E) 50

8. STU is a three digit integer that is divisible by 4. TU could be which of the following?

(A) 22
(B) 26
(C) 32
(D) 45
(E) 50

4.4.3 Positives and Negatives

Facts:
* A positive number multiplied or divided by a positive number yields a positive result.
* A negative number multiplied or divided by a positive number yields a negative result.
* A negative number multiplied or divided by a negative number yields a positive result.

Demonstration Example

Demo: If the product of five integers is negative, how many of these integers could be negative at most?

Because a positive times a positive is a positive, you need at least one negative integer. The other four integers need to result in a positive when multiplied together. Using the rules above, you can arrive at the following combinations:

+ + + + - = -   - - - + + = -   - - - - - = -

From these patterns, you can derive a simple rule: an *even* number of negatives yields a positive result, while an *odd* number of negatives yields a negative result. So, the answer "at most" would be <u>5 negatives</u>.

SAT Example and Technique Application:

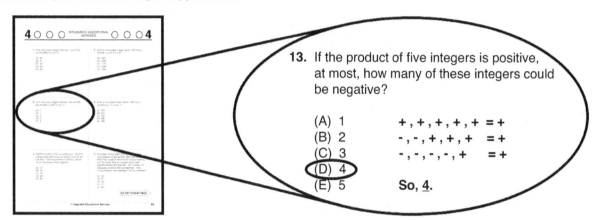

13. If the product of five integers is positive, at most, how many of these integers could be negative?

(A) 1   +, +, +, +, + = +
(B) 2   -, -, +, +, + = +
(C) 3   -, -, -, -, + = +
(D) 4
(E) 5   So, <u>4</u>.

4.4.3 Let's Practice:

1. Ⓐ Ⓑ Ⓒ Ⓓ Ⓔ   6. Ⓐ Ⓑ Ⓒ Ⓓ Ⓔ
2. Ⓐ Ⓑ Ⓒ Ⓓ Ⓔ   7. Ⓐ Ⓑ Ⓒ Ⓓ Ⓔ
3. Ⓐ Ⓑ Ⓒ Ⓓ Ⓔ   8. Ⓐ Ⓑ Ⓒ Ⓓ Ⓔ
4. Ⓐ Ⓑ Ⓒ Ⓓ Ⓔ   9. Ⓐ Ⓑ Ⓒ Ⓓ Ⓔ
5. Ⓐ Ⓑ Ⓒ Ⓓ Ⓔ   10. Ⓐ Ⓑ Ⓒ Ⓓ Ⓔ

1. If the product of four integers is negative, how many of these integers could be negative?

(A) 1 only
(B) 2 only
(C) 3 only
(D) 1 and 2
(E) 1 and 3

2. If the product of four integers is negative, can we determine whether the largest integer is positive or negative?

(A) Yes, Negative.
(B) Yes, Positive.
(C) Yes, Neither, it is 0.
(D) No, It could be either.
(E) No, there is not enough information.

4.5 Exponents

4.5.1

Same Base, Different Exponent	Different Base, Same Exponent
$x^a \cdot x^b = x^{a+b}$	$x^a y^a = (xy)^a$
$x^a/x^b = x^{a-b}$	$x^a/y^a = (x/y)^a$
$(x^a)^b = x^{ab}$	

* Always try to manipulate both sides of the equation so that the bases are the same on both sides.

Demonstration Example

Demo: $16^x = 4^a$ What is x in terms of a?

$$16^x = 4^a \implies 4^{2x} = 4^a \implies \text{so, } 2x = a \implies \underline{x = a/2}$$

SAT Example and Technique Application:

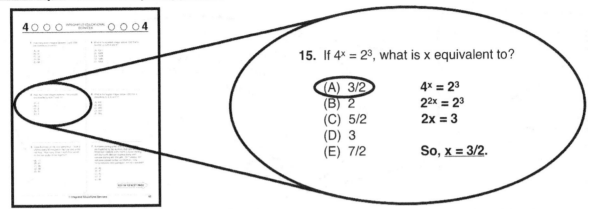

15. If $4^x = 2^3$, what is x equivalent to?

    (A) 3/2
    (B) 2
    (C) 5/2
    (D) 3
    (E) 7/2

$4^x = 2^3$
$2^{2x} = 2^3$
$2x = 3$

So, $\underline{x = 3/2}$.

4.5.1 **Let's Practice:**

1. Ⓐ Ⓑ Ⓒ Ⓓ Ⓔ   6. Ⓐ Ⓑ Ⓒ Ⓓ Ⓔ
2. Ⓐ Ⓑ Ⓒ Ⓓ Ⓔ   7. Ⓐ Ⓑ Ⓒ Ⓓ Ⓔ
3. Ⓐ Ⓑ Ⓒ Ⓓ Ⓔ   8. Ⓐ Ⓑ Ⓒ Ⓓ Ⓔ
4. Ⓐ Ⓑ Ⓒ Ⓓ Ⓔ   9. Ⓐ Ⓑ Ⓒ Ⓓ Ⓔ
5. Ⓐ Ⓑ Ⓒ Ⓓ Ⓔ   10. Ⓐ Ⓑ Ⓒ Ⓓ Ⓔ

1. If $3^x = 27^2$, what does x equal?

    (A) 3
    (B) 4
    (C) 5
    (D) 6
    (E) 7

2. If $16^x = 2^{12}$, what does x equal?

    (A) 0
    (B) 1
    (C) 2
    (D) 3
    (E) 4

3. If $5^9 = 125^x$, what does x equal?

    (A) 4
    (B) 3
    (C) 2
    (D) 1
    (E) 0

4. If $2^x = 4^{12}$, what does x equal?

    (A) 24
    (B) 18
    (C) 16
    (D) 12
    (E) 6

5. If $49^6 = 7^x$, what does x equal?

    (A) 3
    (B) 12
    (C) 30
    (D) 32
    (E) 64

6. If $4^4/64 = x$, what does x equal?

    (A) 0
    (B) 1
    (C) 2
    (D) 3
    (E) 4

GO ON TO NEXT PAGE ▷

4.5.2 Harder exponent questions require manipulation on both sides. Try this:

Demonstration Example

Demo: If $25^y = 64$, what is 5^{y+1} ?

$$25^y = 64 \quad \Rightarrow \quad 5^y = 8 \quad \Rightarrow \quad 5 \cdot 5^y = 5 \cdot 8 \quad \Rightarrow \quad 5^{y+1} = \underline{40}.$$

SAT Example and Technique Application:

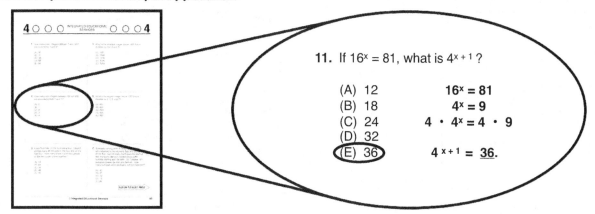

11. If $16^x = 81$, what is 4^{x+1} ?

(A) 12 $16^x = 81$
(B) 18 $4^x = 9$
(C) 24 $4 \cdot 4^x = 4 \cdot 9$
(D) 32
(E) 36 $4^{x+1} = \underline{36}.$

4.5.2 Let's Practice:

1. Ⓐ Ⓑ Ⓒ Ⓓ Ⓔ   6. Ⓐ Ⓑ Ⓒ Ⓓ Ⓔ
2. Ⓐ Ⓑ Ⓒ Ⓓ Ⓔ   7. Ⓐ Ⓑ Ⓒ Ⓓ Ⓔ
3. Ⓐ Ⓑ Ⓒ Ⓓ Ⓔ   8. Ⓐ Ⓑ Ⓒ Ⓓ Ⓔ
4. Ⓐ Ⓑ Ⓒ Ⓓ Ⓔ   9. Ⓐ Ⓑ Ⓒ Ⓓ Ⓔ
5. Ⓐ Ⓑ Ⓒ Ⓓ Ⓔ   10. Ⓐ Ⓑ Ⓒ Ⓓ Ⓔ

1. If $16 = 4^{x+1}$, what does x equal?

(A) 0
(B) 1
(C) 2
(D) 4
(E) 8

2. If $9^x = 25$, what does 3^{x-1} equal?

(A) 4/5
(B) 5/4
(C) 5/3
(D) 2
(E) 3

3. If $3^{5x-3} = 27^4$, what does x equal?

(A) 3
(B) 6
(C) 18
(D) 24
(E) 32

4. If $4^x = 121$, what does 2^{x-3} equal?

(A) 11/2
(B) 9/2
(C) 11/4
(D) 9/4
(E) 11/8

GO ON TO NEXT PAGE ⟩

4.5.3 What happens when there is no easy rule? You have to be creative.

Demonstration Example

Demo: If $2^x + 2^x = 64$, what is x equivalent to?

$2^x + 2^x = 64 \implies (2)2^x = 64 \implies 2^{x+1} = 2^6 \implies x + 2 = 6 \implies x = \underline{4}.$

SAT Example and Technique Application:

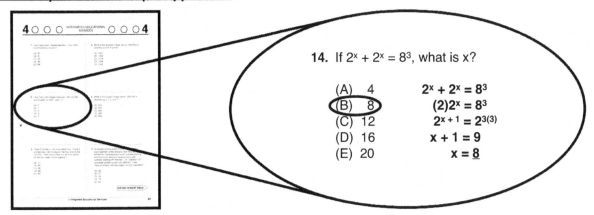

14. If $2^x + 2^x = 8^3$, what is x?

(A) 4
(B) 8
(C) 12
(D) 16
(E) 20

$2^x + 2^x = 8^3$
$(2)2^x = 8^3$
$2^{x+1} = 2^{3(3)}$
$x + 1 = 9$
$x = \underline{8}$

4.5.3 Let's Practice:

1. Ⓐ Ⓑ Ⓒ Ⓓ Ⓔ   6. Ⓐ Ⓑ Ⓒ Ⓓ Ⓔ
2. Ⓐ Ⓑ Ⓒ Ⓓ Ⓔ   7. Ⓐ Ⓑ Ⓒ Ⓓ Ⓔ
3. Ⓐ Ⓑ Ⓒ Ⓓ Ⓔ   8. Ⓐ Ⓑ Ⓒ Ⓓ Ⓔ
4. Ⓐ Ⓑ Ⓒ Ⓓ Ⓔ   9. Ⓐ Ⓑ Ⓒ Ⓓ Ⓔ
5. Ⓐ Ⓑ Ⓒ Ⓓ Ⓔ   10. Ⓐ Ⓑ Ⓒ Ⓓ Ⓔ

1. If $4^x + 4^x = 32$, what is x?

(A) 0
(B) 1
(C) 2
(D) 4
(E) 8

2. If $9^x + 9^x + 9^x = 81^7$, what is x?

(A) 12
(B) 25/2
(C) 13
(D) 27/2
(E) 14

3. If $2^x + 2^x + 2^x + 2^x + 2^x + 2^x + 2^x + 2^x = 256$, what is x?

(A) 3
(B) 4
(C) 5
(D) 6
(E) 7

4. If $5^x + 5^x = 250$, what is x?

(A) 3
(B) 4
(C) 5
(D) 6
(E) 7

GO ON TO NEXT PAGE ⟩

4.5.4 Fractional and Negative Exponents

* When you see a negative exponent, you take the reciprocal of the base.
* When you see a fractional exponent such as 1/2, you take the "denominator" root.
  In other words, $x^{1/2} = \sqrt{x}$, $x^{1/3} = \sqrt[3]{x}$, and so on.

Demonstration Example

Demo: If $x^{2/3} = y^{5/4}$, what is x in terms of y?

$$x^{2/3} = y^{5/4} \implies x^{2/3 \,(3/2)} = y^{5/4 \,(3/2)} \implies \underline{x = y^{15/8}}$$

SAT Example and Technique Application:

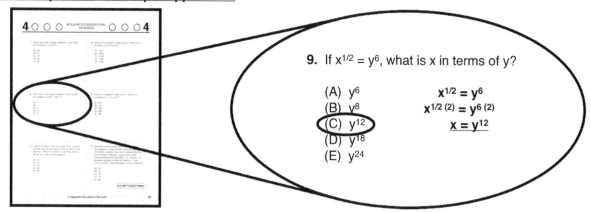

9. If $x^{1/2} = y^6$, what is x in terms of y?

(A) y^6 $x^{1/2} = y^6$
(B) y^8 $x^{1/2\,(2)} = y^{6\,(2)}$
(C) y^{12} $\underline{x = y^{12}}$
(D) y^{18}
(E) y^{24}

4.5.4 Let's Practice:

1. Ⓐ Ⓑ Ⓒ Ⓓ Ⓔ   6. Ⓐ Ⓑ Ⓒ Ⓓ Ⓔ
2. Ⓐ Ⓑ Ⓒ Ⓓ Ⓔ   7. Ⓐ Ⓑ Ⓒ Ⓓ Ⓔ
3. Ⓐ Ⓑ Ⓒ Ⓓ Ⓔ   8. Ⓐ Ⓑ Ⓒ Ⓓ Ⓔ
4. Ⓐ Ⓑ Ⓒ Ⓓ Ⓔ   9. Ⓐ Ⓑ Ⓒ Ⓓ Ⓔ
5. Ⓐ Ⓑ Ⓒ Ⓓ Ⓔ   10. Ⓐ Ⓑ Ⓒ Ⓓ Ⓔ

1. If $x^{-1/3} = y^{-9}$, what is x in terms of y?

(A) y^9
(B) y^{-9}
(C) y^{27}
(D) y^{-27}
(E) y^{81}

2. If $x^{5/2} = y^{10}$, what is x in terms of y?

(A) y^2
(B) y^4
(C) y^6
(D) y^8
(E) y^{20}

3. If $x^{2/3} = y$, what does x^4 equal in terms of y?

(A) $y^{3/2}$
(B) y^3
(C) $y^{5/2}$
(D) $y^{8/3}$
(E) y^6

4. If $x^{-3/2} = 8$, what is the value of x?

(A) 4
(B) 2
(C) 1
(D) 1/4
(E) 1/8

GO ON TO NEXT PAGE ⟩

4 ◯ ◯ ◯

Unauthorized copying or reuse of any part of this page is illegal.

◯ ◯ ◯ 4

4.5.5 **Advanced Exponent Problems**

1. If $8^x \cdot 8^3 = 8^{17}$, what is the value of x?

2. If $7^{2x} = 49a^2$ and $7^x = 8ab$, where a and b are both constants, what is the value of b?

3. $\sqrt{(xy)} = 2$ Find x^4y^4.

4. If $3^x + 3^{x+2} = 880$, what is the value of 3^{x+1}?

5. If $64^x = 4^y$, what is the value of x/y?

6. If $x^4 = y^8z^{16}$ and $y^2z^4 = 3$, what does x equal?

7. If $7^{3a} \cdot 7^{3b} = 49$, what is the value of a + b?

8. If $a^4 = b^{12}$ and a = 5, what is b^{24}/b^{15}?

9. If $(4m)^x = m^{3x}$, where m is a constant, what is the value of m?

10. If $b^x \cdot b^y = b^8$ and $b^x/b^y = b^{12}$, what is the value of x?

GO ON TO NEXT PAGE ⟶

4.6 The Number Line

4.6.1 The Four Zones

* The most important section of the number line is between -1 and 1. There are 4 sections, each with its characteristics such that all numbers within that section share one property. Thus, all you have to do is learn the characteristic for each section.

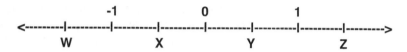

* **Characteristics:**

$W^3 < W < W^2$
$X < X^3 < X^2$
$Y^3 < Y^2 < Y$
$Z < Z^2 < Z^3$

* <u>Important Note</u>: The only characteristic that remains true REGARDLESS of section of the number line is that ADDING a positive number will ALWAYS increase the value of any number.

<u>SAT Example and Technique Application</u>:

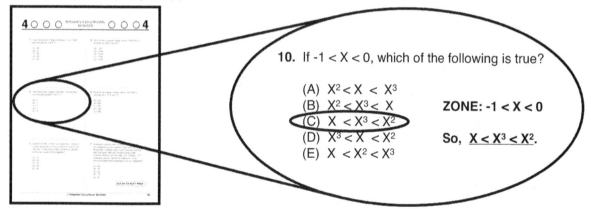

10. If $-1 < X < 0$, which of the following is true?

(A) $X^2 < X < X^3$
(B) $X^2 < X^3 < X$
(C) $X < X^3 < X^2$
(D) $X^3 < X < X^2$
(E) $X < X^2 < X^3$

ZONE: $-1 < X < 0$

So, $X < X^3 < X^2$.

4.6.1 <u>Let's Practice</u>:

1. Ⓐ Ⓑ Ⓒ Ⓓ Ⓔ   6. Ⓐ Ⓑ Ⓒ Ⓓ Ⓔ
2. Ⓐ Ⓑ Ⓒ Ⓓ Ⓔ   7. Ⓐ Ⓑ Ⓒ Ⓓ Ⓔ
3. Ⓐ Ⓑ Ⓒ Ⓓ Ⓔ   8. Ⓐ Ⓑ Ⓒ Ⓓ Ⓔ
4. Ⓐ Ⓑ Ⓒ Ⓓ Ⓔ   9. Ⓐ Ⓑ Ⓒ Ⓓ Ⓔ
5. Ⓐ Ⓑ Ⓒ Ⓓ Ⓔ   10. Ⓐ Ⓑ Ⓒ Ⓓ Ⓔ

1. If $0 < X < 1$, which of the following is true?

(A) $X^2 < X < \sqrt{X}$
(B) $X^2 < \sqrt{X} < X$
(C) $X < \sqrt{X} < X^2$
(D) $\sqrt{X} < X < X^2$
(E) $X < X^2 < \sqrt{X}$

2. If $X < -1$, which of the following is true?

(A) $X < 2X < X^2$
(B) $X < X + 1 < 2X$
(C) $X + 1 < X < X^2$
(D) $X < X^2 < X + 1$
(E) $X < X + 1 < X^2$

GO ON TO NEXT PAGE ⟩

4 ◯ ◯ ◯

Unauthorized copying or
reuse of any part of this
page is illegal.

◯ ◯ ◯ 4

4.6.2 **Magnitudes**

* The expression |A - B| is to be read as the distance between points A and B.
* The expression |X + Y| is to be read as the distance between points X and -Y (|X - (-Y)|).

$$< \text{--------} | \text{--------} | \text{--------} | \text{--------} | \text{--------} | \text{--------} | \text{--------} | \text{--------} | \text{--------} | \text{--------} >$$
$$\quad\quad a \quad\quad\quad b \quad\quad\quad 0 \quad\; c \quad\quad\quad\quad\quad d$$

* Considering the number line above, which of the following expressions has the greatest value?

(A) | a - b |
(B) | c - a |
(C) | b - c |
(D) | a + c |
(E) | a - d |

* <u>Note</u>: It is important (and faster!) to analyze these expressions
as the "distance between the points" and to compare these
distances, rather than plugging numbers into the expressions.

4.6.2 **Let's Practice:** **Use the following number line to answer the questions below.**

$$< \text{--------} | \text{--------} | \text{--------} | \text{--------} | \text{--------} | \text{--------} | \text{--------} | \text{--------} | \text{--------} | \text{--------} >$$
$$\quad\quad a \quad\quad\quad\quad\quad\; b \quad 0 \quad\quad\quad c \quad\quad\quad d$$

1. Ⓐ Ⓑ Ⓒ Ⓓ Ⓔ   6. Ⓐ Ⓑ Ⓒ Ⓓ Ⓔ
2. Ⓐ Ⓑ Ⓒ Ⓓ Ⓔ   7. Ⓐ Ⓑ Ⓒ Ⓓ Ⓔ
3. Ⓐ Ⓑ Ⓒ Ⓓ Ⓔ   8. Ⓐ Ⓑ Ⓒ Ⓓ Ⓔ
4. Ⓐ Ⓑ Ⓒ Ⓓ Ⓔ   9. Ⓐ Ⓑ Ⓒ Ⓓ Ⓔ
5. Ⓐ Ⓑ Ⓒ Ⓓ Ⓔ   10. Ⓐ Ⓑ Ⓒ Ⓓ Ⓔ

1. Which expression has the greatest value?

(A) |a - b|
(B) |a - c|
(C) |b + c|
(D) |a + b|
(E) |a + c|

2. Which expression has the least value?

(A) ab
(B) ac
(C) cd
(D) bc
(E) ad

4.6.3 **Equally-Spaced Tick Marks**

* To find C on the following number line on which all tick marks are equally-spaced, determine the
spacing between tick marks by counting the spaces between two known numbers and then dividing.

$$< \text{--------} | \text{--------} | \text{--------} | \text{--------} | \text{--------} | \text{--------} | \text{--------} | \text{--------} | \text{--------} | \text{--------} >$$
$$\quad\quad 5 \quad\quad\quad 6 \quad\quad\quad 7 \quad\quad\quad\quad\quad C$$

* Because there are two spaces between 5 and 6, we can determine that the value of each space is
(6-5)/2 = 1/2. There are three spaces between 7 and C, so C = 7 + 3(1/2) = <u>8 1/2</u>.

4.6.3 **Equally Spaced Tick Marks Problems**   (Solve for $\underline{x}$ unless otherwise noted.)

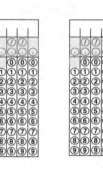

---

1. < --- | --- | --- | --- | --- | --- | --- | --- | --- | --- >
        2   9   x

---

2. < --- | --- | --- | --- | --- | --- | --- | --- | --- | --- >
        -5   x   30

---

3. < --- | --- | --- | --- | --- | --- | --- | --- | --- | --- >
        -1   x   1

---

4. < --- | --- | --- | --- | --- | --- | --- | --- | --- | --- >
        0   a   b   c

* If all tick marks are equally spaced and
   b - a = 12, find c.

---

5. < --- | --- | --- | --- | --- | --- | --- | --- | --- | --- >
        p   0   q   r

* If all tick marks are equally spaced and
   p + r = 5, find q.

---

6. < --- | --- | --- | --- | --- | --- | --- | --- | --- | --- >
        x   1/2   4/5

---

7. < --- | --- | --- | --- | --- | --- | --- | --- | --- | --- >
        -2/5   0   x

---

8. < --- | --- | --- | --- | --- | --- | --- | --- | --- | --- >
        7/9   x   1

---

9. < --- | --- | --- | --- | --- | --- | --- | --- | --- | --- >
        a   b   0   c

* If all tick marks are equally spaced and
   a + b = -2, find c.

GO ON TO NEXT PAGE ▷

4.7 Absolute Value

All absolute value functions are distance functions which measure the distance between two points. For example, |x - y| = 12 means that the distance between x and y is 12.

Demonstration Examples

* For absolute value "equalities"...

Demo 1: |x - 3| = 8 Find x.

* Remember, this is a measure of distance. The easiest way to read the statement above is... "What numbers are _8_ away from _3_ ?" On a number line, this is 3 ± 8, _-5_ and _11_.

* And remember, first, that order doesn't matter, so |x - 8| is the same as |8 - x|. Second, if the statement has a positive sign, just change the number to its opposite.

|3 + x| = 8   "What numbers are _8_ away from <u>Negative 3</u>?"   <u>-11 and 5</u>

Demo 2: |3x - 21| = 9 Find x.

* It is important to note that it is impossible to apply this technique unless the variable is only multiplied by one. In this case, we must first divide everything by _3_, so that we just have x.

|3x - 21| = 9   --> |x - 7| = 3   -->   "What numbers are _3_ away from _7_?"   <u>4 and 10</u>

---

* For absolute value "inequalities"...

Demo 3: |x - 6| < 2 Find all values of x that satisfy the inequality.

* These inequality problems are solved the same way as the equalities. You just have to remember the following...

< means "in between"   > means "outside of"

"What numbers are 2 away from 6?"   <u>4 and 8</u>   < means "in between". Therefore, <u>4 < x < 8</u>.

SAT Example and Technique Application:

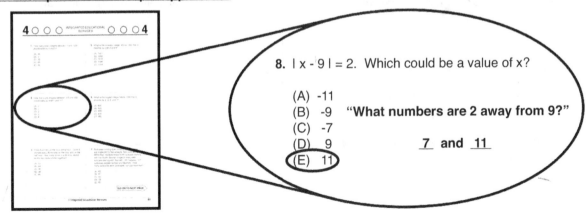

8. | x - 9 | = 2. Which could be a value of x?

(A) -11
(B) -9   "What numbers are 2 away from 9?"
(C) -7
(D) 9   _7_ and _11_
(E) 11

4.7 **Let's Practice:**

1. Ⓐ Ⓑ Ⓒ Ⓓ Ⓔ   6. Ⓐ Ⓑ Ⓒ Ⓓ Ⓔ
2. Ⓐ Ⓑ Ⓒ Ⓓ Ⓔ   7. Ⓐ Ⓑ Ⓒ Ⓓ Ⓔ
3. Ⓐ Ⓑ Ⓒ Ⓓ Ⓔ   8. Ⓐ Ⓑ Ⓒ Ⓓ Ⓔ
4. Ⓐ Ⓑ Ⓒ Ⓓ Ⓔ   9. Ⓐ Ⓑ Ⓒ Ⓓ Ⓔ
5. Ⓐ Ⓑ Ⓒ Ⓓ Ⓔ   10. Ⓐ Ⓑ Ⓒ Ⓓ Ⓔ

1. $|10 - x| = 3$. Which could be a value of x?

(A) -13
(B) -7
(C) 7
(D) 10
(E) 12

2. $|x + 5| = 11$. Which could be a value of x?

(A) -16
(B) -6
(C) -5
(D) 5
(E) 11

3. $|12 + x| = 32$. Which could be a value of x?

(A) -20
(B) -12
(C) -10
(D) 12
(E) 20

4. $|93 - 3x| = 12$. Which could be a value of x?

(A) -35
(B) -21
(C) 4
(D) 27
(E) 31

5. $|2x - 6| = 4$. Which could be a value of x?

(A) 3
(B) 4
(C) 5
(D) 6
(E) 10

6. $|x - 3| < 7$. Which could be a value of x?

(A) -13
(B) -7
(C) 7
(D) 10
(E) 12

7. $|4 - x| \leq 12$. Which could be a value of x?

(A) -12
(B) -8
(C) 20
(D) 24
(E) 32

8. $|9 + 2x| > 2$. Which could be a value of x?

(A) -6
(B) -5.5
(C) -5
(D) -4.5
(E) -4

9. $|x - 23| > 12$. Which represents all values of x?

(A) -35 < x < -11
(B) -11 < x < 11
(C) 11 < x < 35
(D) x < -35 or x > -11
(E) x < 11 or x > 35

10. If $|x - 8| < |x - 12|$, which of the following represents all values of x that satisfy the inequality?

(A) x < 12
(B) x > 12
(C) x < 10
(D) x > 10
(E) x > 8

GO ON TO NEXT PAGE ⟩

4.8 Roots

Demonstration Examples

Demo 1: If $(x - 2)(x + 3) = 0$. What are all possible solutions for x?

* **2 and -3 are the zeros of this equation, so they are the possible solutions.**

Demo 2: If $(x - 2)(x + 3) < 0$. What are all possible solutions for x?

* **As before, < means "in between" and > means "outside of," So, $-3 < x < 2$.**

SAT Example and Technique Application:

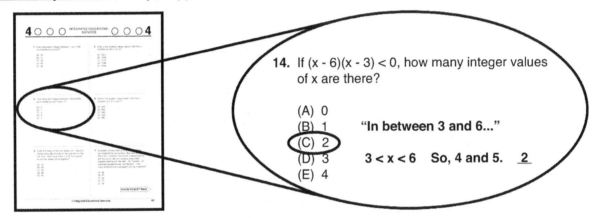

14. If $(x - 6)(x - 3) < 0$, how many integer values of x are there?

(A) 0
(B) 1
(C) 2
(D) 3
(E) 4

"In between 3 and 6..."

$3 < x < 6$ So, 4 and 5.   <u>2</u>

4.8 Let's Practice:

1. Ⓐ Ⓑ Ⓒ Ⓓ Ⓔ   6. Ⓐ Ⓑ Ⓒ Ⓓ Ⓔ
2. Ⓐ Ⓑ Ⓒ Ⓓ Ⓔ   7. Ⓐ Ⓑ Ⓒ Ⓓ Ⓔ
3. Ⓐ Ⓑ Ⓒ Ⓓ Ⓔ   8. Ⓐ Ⓑ Ⓒ Ⓓ Ⓔ
4. Ⓐ Ⓑ Ⓒ Ⓓ Ⓔ   9. Ⓐ Ⓑ Ⓒ Ⓓ Ⓔ
5. Ⓐ Ⓑ Ⓒ Ⓓ Ⓔ   10. Ⓐ Ⓑ Ⓒ Ⓓ Ⓔ

1. What is the minimum positive integer value of x for which $(x - 2)(x - 3) > 0$?

(A) 0
(B) 1
(C) 2
(D) 3
(E) 4

2. What is the maximum negative integer value of x for which $(x + 1)(x - 3) > 0$?

(A) -2
(B) -1
(C) 0
(D) 1
(E) 2

3. If x is an integer, find x when $(3x - 2)(5x + 3) < 0$.

(A) -1
(B) 0
(C) 1
(D) 2
(E) 3

GO ON TO NEXT PAGE ▷

CHAPTER 4: CHALLENGE QUESTIONS   Student-Produced Responses

4.1

1. How many positive integers less than 1000 are divisible by 2, 3, 7, or 14?

4.4

2. The sum of all of the factors of the squares of two prime numbers is 44. How many factors does the product of these two numbers have?

3. Suppose that the product of 7 numbers is divided by the square of the product of the same 7 numbers. If the answer is negative, how many of these numbers could be negative at most?

4.5

4. If $(4^x + 4^x + 4^x + 4^x)^4 = 256^{3x}$, what is the value of x?

5. If $125^{2x}/729m^9 = (81m)(9m^8)$, what is the value of 5^x when m is equal to 2?

4.7

6. If $|x - 8| < |x - 12|$, what is the smallest positive integer value of x that satisfies the inequality?

4.8

7. If $y > 5x^2 - 4x - 1$, what is one positive value of x that satisfies the inequality and that is also a fractional perfect square?

8. What is the lowest value of x that satisfies the inequality $y \geq (3x - 13)(4x - 17)$?

GO ON TO NEXT PAGE ▷

Multiple-Choice	Student-Produced Responses

1 Ⓐ Ⓑ Ⓒ Ⓓ Ⓔ
2 Ⓐ Ⓑ Ⓒ Ⓓ Ⓔ
3 Ⓐ Ⓑ Ⓒ Ⓓ Ⓔ
4 Ⓐ Ⓑ Ⓒ Ⓓ Ⓔ
5 Ⓐ Ⓑ Ⓒ Ⓓ Ⓔ
Ⓐ Ⓑ Ⓒ Ⓓ Ⓔ
Ⓐ Ⓑ Ⓒ Ⓓ Ⓔ
Ⓐ Ⓑ Ⓒ Ⓓ Ⓔ
Ⓐ Ⓑ Ⓒ Ⓓ Ⓔ
Ⓐ Ⓑ Ⓒ Ⓓ Ⓔ

CHAPTER 4 REVIEW

6, 7, 8, 9 grids and an additional grid (student-produced response bubble grids)

---

1. If $3^x = 25$, what is 3^{x+2}?

(A) 175
(B) 180
(C) 200
(D) 225
(E) 250

2. If $(m^n)^2 = m^3/m^n$, what is the value of n?

(A) 1
(B) 2
(C) 5
(D) 8
(E) 10

3. K is a number that, when divided by 7, gives a remainder of 3. What is the remainder when 4k is divided by 7?

(A) 3
(B) 4
(C) 5
(D) 6
(E) 7

4. What time is it 65 hours after 3:00pm?

(A) 7:00 AM
(B) 8:00 AM
(C) 7:00 PM
(D) 8:00 PM
(E) 9:00 PM

5. $|x + 3| = 5$. What is the lower value of x?

(A) -8
(B) -5
(C) -3
(D) 3
(E) 8

6. A spinner on a game board has 4 numbered regions: 25, 50, 75, and 100 (in that order). If the spinner starts on 75 and is spun so hard that it passes through 70 regions before stopping, what region does it stop on?

7. Suppose that $|8 + x| \le 10$. What is the greatest possible value for x?

8. How many positive integers less than 200 are divisible by either 3 or 4?

9. What number between 60 and 70 gives a remainder of 2 when divided by 13?

10. For an outer space mission, a shuttle needs a precise combination of weight, speed, and pressure to function optimally. For this reason, an ideal "space weight" has been determined by NASA scientists to dictate whether or not an astronaut can board a shuttle. The ideal space weight is set at 130 lbs., and no astronaut who is more than 40 lbs. away from this ideal weight may board a shuttle. Write an absolute value inequality to determine whether or not an astronaut's weight, w, is within the proper weight range. *(Write your answer here. Do not grid.)*

GO ON TO NEXT PAGE ⟶

4 ◯ ◯ ◯

Unauthorized copying or
reuse of any part of this
page is illegal.

◯ ◯ ◯ 4

Multiple-Choice

Student-Produced Responses

CHAPTER

1 - 4

CUMULATIVE
REVIEW

1 Ⓐ Ⓑ Ⓒ Ⓓ Ⓔ
2 Ⓐ Ⓑ Ⓒ Ⓓ Ⓔ
3 Ⓐ Ⓑ Ⓒ Ⓓ Ⓔ
4 Ⓐ Ⓑ Ⓒ Ⓓ Ⓔ
  Ⓐ Ⓑ Ⓒ Ⓓ Ⓔ
  Ⓐ Ⓑ Ⓒ Ⓓ Ⓔ
  Ⓐ Ⓑ Ⓒ Ⓓ Ⓔ
  Ⓐ Ⓑ Ⓒ Ⓓ Ⓔ
  Ⓐ Ⓑ Ⓒ Ⓓ Ⓔ
  Ⓐ Ⓑ Ⓒ Ⓓ Ⓔ

5  6  7  8 9

1. 5% of 8% of a number is what percent of 4% of the same number?

    (A) 2%
    (B) 6%
    (C) 10%
    (D) 12%
    (E) 20%

2. If $6^x/36^{x/4} = 216^{x+1}$, what is the value of x?

    (A) 6/5
    (B) 5/6
    (C) 1/3
    (D) -1/3
    (E) -6/5

3. The weight of two wolf cubs and 1 bear cub is 140 pounds and the weight of one wolf cub and 2 bear cubs is 110 pounds. What is the combined weight of one bear cub and one wolf cub?

    (A) 80
    (B) 82.4
    (C) 83.3
    (D) 84.1
    (E) 90.3

4. The average of F, G, H, and J is 40. When a fifth number X is added, the new average is 36. What is the value of X?

    (A) 20
    (B) 18
    (C) 16
    (D) 12
    (E) 10

5. The average score on a test in a class of A students is 75. The average score on the same test in a class of B students is 95. If the combined average for the two classes is 90, what is the value of A/B?

6. The relationship between a jockey's weight and a horse's speed is inversely proportional. So, the lighter the weight of a jockey, the faster the jockey's horse can run. If a horse can run 40 mph with a jockey who weighs 90 pounds, how much faster can the horse run with a jockey who weighs 60 pounds?

7. 11 times a number is 8 more than 3 times the same number. What is the number?

8. At a Japanese garden, the ratio of red koi fish to normal koi fish in a pond is 3:5. If there are 40 koi fish in the pond all together, how many of the koi fish are red?

9. The ratio of white tea tins to green tea tins sold at a tea store is 5 to 4. White tea costs $12 per tin, whereas green tea costs $7 per tin. If the store made $792 selling the two types of tea, how many tins of white tea were sold?

10. In order to board a ride at an amusement park, a person can be no shorter than 5' and no taller than 6'6". Write an inequality that determines whether a person with height x can board the ride? *(Write your answer here. Do not grid.)*

STOP

Chapter 5: Rates of Change

4 Sections
50 Practice Questions

5.1 Split Rate Formula

Whenever you face a problem in which there are two different prices or rates depending on the quantity bought or on the time used for a service (like phone calls), you should use the following formula:

$$C_n = R_1Q_1 + R_2(Q_n - Q_1) \text{ , where C = cost, R = rate, and Q = quantity.}$$

Demonstration Example

Demo: A store charges $5 per shirt for the first shirt and $3 per shirt after the first shirt purchased. What is the total cost if you buy 9 shirts?

$$C_9 = 5(1) + 3(9 - 1) \qquad C_9 = 5 + 24 \qquad C_9 = \underline{29}$$

SAT Example and Technique Application:

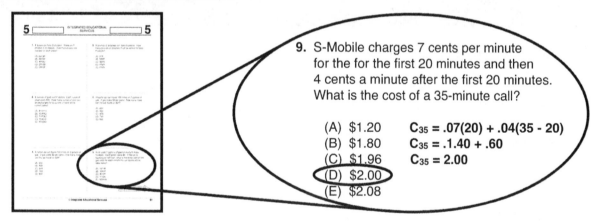

9. S-Mobile charges 7 cents per minute for the for the first 20 minutes and then 4 cents a minute after the first 20 minutes. What is the cost of a 35-minute call?

(A) $1.20 $C_{35} = .07(20) + .04(35 - 20)$
(B) $1.80 $C_{35} = .1.40 + .60$
(C) $1.96 $C_{35} = 2.00$
(D) $2.00
(E) $2.08

5.1 Let's Practice:

1. Ⓐ Ⓑ Ⓒ Ⓓ Ⓔ   6. Ⓐ Ⓑ Ⓒ Ⓓ Ⓔ
2. Ⓐ Ⓑ Ⓒ Ⓓ Ⓔ   7. Ⓐ Ⓑ Ⓒ Ⓓ Ⓔ
3. Ⓐ Ⓑ Ⓒ Ⓓ Ⓔ   8. Ⓐ Ⓑ Ⓒ Ⓓ Ⓔ
4. Ⓐ Ⓑ Ⓒ Ⓓ Ⓔ   9. Ⓐ Ⓑ Ⓒ Ⓓ Ⓔ
5. Ⓐ Ⓑ Ⓒ Ⓓ Ⓔ   10. Ⓐ Ⓑ Ⓒ Ⓓ Ⓔ

1. The Horizon Phone Company charges one dollar for any time up to one minute, then 3 cents per minute after the first minute. What is the cost of a 12-minute call?

(A) $1.33
(B) $1.36
(C) $1.39
(D) $1.42
(E) $1.45

GO ON TO NEXT PAGE ⇒

5

5

Unauthorized copying or reuse of any part of this page is illegal.

2. Given that S-Mobile charges 7 cents a minute for the first 20 minutes and 4 cents per minute after the first 20 minutes and given that the Horizon Phone Company charges one dollar for any time up to one minute and 3 cents per minute after the first minute, how long would a phone call last that costs the same amount through either provider?

(A) 40
(B) 37
(C) 36
(D) 33
(E) 32

3. Stamps Inc. charges $3 per oz. for the first 8 oz. of goods shipped and $2 per oz. thereafter. If IES2400 pays $112 to ship out its SAT Verbal Primers, how many ounces do the books shipped weigh in total?

(A) 50
(B) 52
(C) 56
(D) 60
(E) 64

4. Stamps Publishing charges $15 for printing a first book, $5 per book for the next 19 books, and $2 per book thereafter. How much would Stamps Publishing charge for 35 copies of the IES2400 SAT Math Technique Workbook?

(A) $175
(B) $165
(C) $155
(D) $150
(E) $140

5.2 Distance = Speed x Time

5.2.1 Basic Distance = Speed x Time Questions

The basic distance equation, Distance = Rate x Time, should be used when the SAT gives you two of the quantities and wants you to solve for the third.

5.2.1 Let's Practice:

1. Ⓐ Ⓑ Ⓒ Ⓓ Ⓔ   6. Ⓐ Ⓑ Ⓒ Ⓓ Ⓔ
2. Ⓐ Ⓑ Ⓒ Ⓓ Ⓔ   7. Ⓐ Ⓑ Ⓒ Ⓓ Ⓔ
3. Ⓐ Ⓑ Ⓒ Ⓓ Ⓔ   8. Ⓐ Ⓑ Ⓒ Ⓓ Ⓔ
4. Ⓐ Ⓑ Ⓒ Ⓓ Ⓔ   9. Ⓐ Ⓑ Ⓒ Ⓓ Ⓔ
5. Ⓐ Ⓑ Ⓒ Ⓓ Ⓔ   10. Ⓐ Ⓑ Ⓒ Ⓓ Ⓔ

1. How far do you travel when you drive at 40mph for 3 hours?

(A) 13.3 miles
(B) 40.0 miles
(C) 53.3 miles
(D) 80.0 miles
(E) 120.0 miles

2. How long (in minutes) does it take to travel 30 miles at 75 mph?

(A) 20 minutes
(B) 24 minutes
(C) 28 minutes
(D) 32 minutes
(E) 36 minutes

3. How fast (in mph) must you drive in order to travel 80 miles in 1 hour and 20 minutes?

(A) 40 mph
(B) 48 mph
(C) 60 mph
(D) 64 mph
(E) 72 mph

GO ON TO NEXT PAGE

5.2.2 Average Speed

Total Distance = Average Speed X Total Time

Average speed is NOT the mean of the given speeds. It is related to total time and total distance. Finding the mean of the speeds will give you the wrong answer. Instead, find the total time and the total distance to get the average speed.

Demonstration Example

Demo: Pete drove for 3 hours at 40mph, then for 5 hours at 50mph, then for 2 hours at 60mph. What was his average speed for the entire trip?

Distance	Speed	Time
120	40	3
250	50	5
120	60	2
490	X = <u>49</u>	10

* **Make a table with columns for Distance, Speed, and Time, as well as a Totals row on the bottom. Use the table to solve for the Average Speed.**

5.2.2 Let's Practice:

1. Ⓐ Ⓑ Ⓒ Ⓓ Ⓔ   6. Ⓐ Ⓑ Ⓒ Ⓓ Ⓔ
2. Ⓐ Ⓑ Ⓒ Ⓓ Ⓔ   7. Ⓐ Ⓑ Ⓒ Ⓓ Ⓔ
3. Ⓐ Ⓑ Ⓒ Ⓓ Ⓔ   8. Ⓐ Ⓑ Ⓒ Ⓓ Ⓔ
4. Ⓐ Ⓑ Ⓒ Ⓓ Ⓔ   9. Ⓐ Ⓑ Ⓒ Ⓓ Ⓔ
5. Ⓐ Ⓑ Ⓒ Ⓓ Ⓔ   10. Ⓐ Ⓑ Ⓒ Ⓓ Ⓔ

1. Sue drove for 10 miles at 40 mph, then for 6 miles at 60 mph, and finally four miles at 80 mph. What was her average speed for the entire trip?

(A) 48 mph
(B) 50 mph
(C) 54 mph
(D) 60 mph
(E) 62 mph

2. Frank drove from home to work at 40 mph and then drove back home at 60 mph without stopping. What was his average speed for the entire trip?

(A) 48 mph
(B) 50 mph
(C) 54 mph
(D) 60 mph
(E) 62 mph

GO ON TO NEXT PAGE

5.2.3 IES2400 Average Speed Formula

IES2400 Average Speed Formula: Average Speed = $\dfrac{2\,(S_1 \times S_2)}{S_1 + S_2}$

* This formula is used when the distance that is being covered remains constant.

Demonstration Example

Demo: Jack travels from home to work at 40 mph and returns from work to home at 60 mph, taking 3 hours total for the roundtrip and not making any stops. How long does it take Jack to travel to work?

Distance	Speed	Time
d	40 mph	
d	60 mph	
	IES Avg. Sp. = $\dfrac{2(40 \times 60)}{(40 + 60)}$	3 hours

So...

Distance	Speed	Time
d = 72 miles	40 mph	1.8 hours
d = 72 miles	60 mph	1.2 hours
144 miles	48 mph	3 hours

* It takes Jack **1.8 hours** to drive to work.

5.2.3 Let's Practice:

1. Ⓐ Ⓑ Ⓒ Ⓓ Ⓔ   6. Ⓐ Ⓑ Ⓒ Ⓓ Ⓔ
2. Ⓐ Ⓑ Ⓒ Ⓓ Ⓔ   7. Ⓐ Ⓑ Ⓒ Ⓓ Ⓔ
3. Ⓐ Ⓑ Ⓒ Ⓓ Ⓔ   8. Ⓐ Ⓑ Ⓒ Ⓓ Ⓔ
4. Ⓐ Ⓑ Ⓒ Ⓓ Ⓔ   9. Ⓐ Ⓑ Ⓒ Ⓓ Ⓔ
5. Ⓐ Ⓑ Ⓒ Ⓓ Ⓔ   10. Ⓐ Ⓑ Ⓒ Ⓓ Ⓔ

1. Steve travels from NYC to Boston at 90 mph and returns to NYC at 60 mph taking 6 hours total for the round-trip without stopping. What is the distance between NYC and Boston?

(A) 192 mi.
(B) 208 mi.
(C) 216 mi.
(D) 224 mi.
(E) 272 mi.

2. Based on the information in the previous question, how long does it take Steve to travel from Boston back to NYC?

(A) 1.2 hours
(B) 1.8 hours
(C) 2.4 hours
(D) 3.6 hours
(E) 4.8 hours

GO ON TO NEXT PAGE

3. Jane drives from Alabama to Arkansas at 30 mph, then back to Alabama at 50 mph without stopping. If the round-trip journey takes 4 hours, how much longer does it take Jane to drive to Arkansas than it takes her to drive back to Alabama?

(A) 1.0 hours
(B) 1.5 hours
(C) 2.0 hours
(D) 2.5 hours
(E) 3.0 hours

4. Paul walks from Point A to Point B at an average speed of 5 miles per hour and runs back from Point B to Point A at an average speed of 20 miles per hour. If Paul does not stop to rest, what is his average speed overall?

(A) 6 mph
(B) 8 mph
(C) 10 mph
(D) 11 mph
(E) 12 mph

5. Melissa reads 48 pages per hour before lunch, but only 16 pages per hour after lunch. If she reads the same number of pages in the morning as in the afternoon, how many pages will she have read after 4 hours of total reading time?

(A) 80 pages
(B) 88 pages
(C) 92 pages
(D) 96 pages
(E) 112 pages

6. A plane travels with headwinds at 300 miles per hour from NY to Paris. The return flight with tailwinds is undertaken at 700 mph. If the total flight time round-trip is 16 hours, how long does it take to go from NY to Paris?

(A) 4 hours, 48 minutes
(B) 11 hours, 12 minutes
(C) 11 hours, 20 minutes
(D) 13 hours, 20 minutes
(E) 13 hours, 33 minutes

7. Frank wants to average 30 mph on his trip from Florida to Memphis. If he only averages 20 mph for the first half of the distance, what should he average for the second half of the distance in order to meet his goal?

(A) 60 mph
(B) 56 mph
(C) 50 mph
(D) 42 mph
(E) 40 mph

GO ON TO NEXT PAGE

5
Unauthorized copying or
reuse of any part of this
page is illegal.
5

5.3 Working Rates

Demonstration Examples

Demo 1: Mike can build a wall in 4 hours. John can build the same wall in 5 hours.
Working together, how long will it take Mike and John to build this wall?

$$\frac{x}{4} + \frac{x}{5} = 1 \qquad \textbf{Multiply by 20.} \qquad 5x + 4x = 20 \qquad 9x = 20 \qquad x = \frac{20}{9} \text{ hours}$$

Demo 2: How would you modify the equation above for 2 walls?

Change the 1 to a 2 and answer the question the same way.

SAT Example and Technique Application:

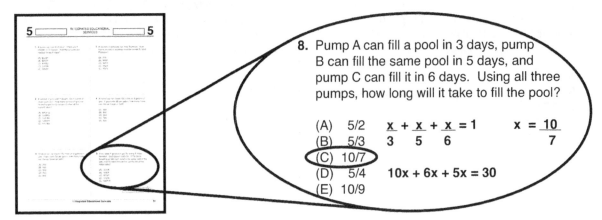

8. Pump A can fill a pool in 3 days, pump
    B can fill the same pool in 5 days, and
    pump C can fill it in 6 days. Using all three
    pumps, how long will it take to fill the pool?

    (A) 5/2 $\qquad \frac{x}{3} + \frac{x}{5} + \frac{x}{6} = 1 \qquad x = \frac{10}{7}$
    (B) 5/3
    (C) 10/7
    (D) 5/4 $\qquad 10x + 6x + 5x = 30$
    (E) 10/9

5.3 Let's Practice:

1. Ⓐ Ⓑ Ⓒ Ⓓ Ⓔ   6. Ⓐ Ⓑ Ⓒ Ⓓ Ⓔ
2. Ⓐ Ⓑ Ⓒ Ⓓ Ⓔ   7. Ⓐ Ⓑ Ⓒ Ⓓ Ⓔ
3. Ⓐ Ⓑ Ⓒ Ⓓ Ⓔ   8. Ⓐ Ⓑ Ⓒ Ⓓ Ⓔ
4. Ⓐ Ⓑ Ⓒ Ⓓ Ⓔ   9. Ⓐ Ⓑ Ⓒ Ⓓ Ⓔ
5. Ⓐ Ⓑ Ⓒ Ⓓ Ⓔ   10. Ⓐ Ⓑ Ⓒ Ⓓ Ⓔ

1. Jack can paint a wall in 20 minutes. Peter can
    paint the same wall in 40 minutes. Working
    together, how many minutes would it take Jack
    and Peter to paint the wall?

    (A) 10/3
    (B) 4
    (C) 5
    (D) 20/3
    (E) 40/3

2. Mary can paint a wall in 1 hour. Simon can
    paint the same wall in 2 hours. When they are
    working together, what fraction of the
    completed wall will Mary paint?

    (A) 1/3
    (B) 1/2
    (C) 5/8
    (D) 2/3
    (E) 3/4

GO ON TO NEXT PAGE

5

5

Unauthorized copying or
reuse of any part of this
page is illegal.

3. The hot water faucet can fill a bathtub in 12 minutes. The cold faucet can fill the same bathtub in 10 minutes. If both faucets are opened, what fraction of the full bath will be filled with hot water?

(A) 5/11
(B) 6/11
(C) 13/5
(D) 18/5
(E) 21/5

4. If the cold water faucet can fill a tub in 20 minutes and the hot water faucet can fill it in 30 minutes, how long will it take to fill half of the tub with both faucets open?

(A) 6 min.
(B) 8 min.
(C) 10 min.
(D) 11 min.
(E) 12 min.

5. If a drain can empty a full tub in 40 minutes and a faucet can fill the tub in 30 minutes, how long will it take to fill the tub when the drain is open?

(A) 90 min.
(B) 100 min.
(C) 120 min.
(D) 130 min.
(E) 150 min.

6. Angela can write a 200-page book in three days, Beth can write a book with the same number of pages in five days. How many days will it take the two of them working together to write 40 200-page books?

(A) 65 days
(B) 75 days
(C) 80 days
(D) 85 days
(E) 90 days

5.4 Multivariable Questions

Option 1: **Dimensional Analysis**

Demonstration Example

Demo: X boxes of pencils contain Y pencils and each pencil costs G dollars. How many boxes can be bought with K dollars?

  A. **Write each variable statement as a fraction:**

$$\frac{X\ Boxes}{Y\ Pencils} \times \frac{1\ Pencil}{G\ Dollars} \times K\ Dollars$$

  B. **See what the question wants (Boxes), and make sure that the units cancel appropriately:**

$$\frac{X\ Boxes}{Y\ \cancel{Pencils}} \times \frac{1\ \cancel{Pencil}}{G\ \cancel{Dollars}} \times K\ \cancel{Dollars}$$

  C. **If the units are correct and no fractions need to be flipped, multiply across for your answer:**

$$\frac{XK}{YG}\ Boxes$$

5 ☐☐☐☐☐☐☐☐

Unauthorized copying or
reuse of any part of this
page is illegal.

☐☐☐☐☐☐☐☐ 5

Option 2: **The Chart**

Demonstration Example

Demo: X boxes of pencils contain Y pencils and each pencil costs G dollars. How
many boxes can be bought with K dollars?

 A. **Underline all the variables: (X boxes, Y pencils, G dollars, K dollars)**

 B. **Circle the target: (How many boxes?)**

 C. **Make a table with as many columns as units: (In this case, 3 columns.)**

 D. **Insert the given variables:**

Boxes	Pencils	Dollars
X	Y	
	1	G
		1
?		K

 E. **Before doing any work on the table, make sure that all the given
information is correctly entered. Notice we have added a row
between G and K. The reason for this is that, in order to convert
one variable into another variable, we always go through 1.**

 F. **Work your way down the table by multiplying or dividing until you
reach the target. Use the following rule:** <u>**If a variable is on top of
the 1, divide by that variable. If a variable is below the 1, you
multiply by that variable.**</u>

 G. **This rule applies to the question below:**

Boxes	Pencils	Dollars
X	Y	
X/Y	1	G
X/YG		1
<u>XK/YG</u>		K

5 [   ]
Unauthorized copying or
reuse of any part of this
page is illegal.
[   ] **5**

SAT Example and Technique Application:

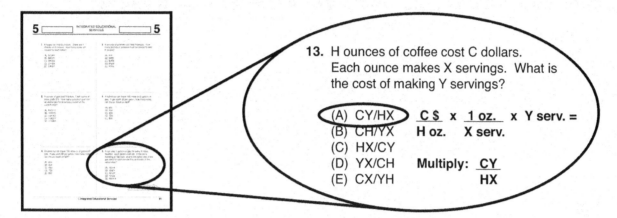

13. H ounces of coffee cost C dollars. Each ounce makes X servings. What is the cost of making Y servings?

(A) CY/HX $\dfrac{C\ \$}{H\ oz.}$ x $\dfrac{1\ oz.}{X\ serv.}$ x Y serv. =

(B) CH/YX

(C) HX/CY

(D) YX/CH   Multiply: $\dfrac{CY}{HX}$

(E) CX/YH

5.4 **Let's Practice:**

1. (A)(B)(C)(D)(E)   6. (A)(B)(C)(D)(E)
2. (A)(B)(C)(D)(E)   7. (A)(B)(C)(D)(E)
3. (A)(B)(C)(D)(E)   8. (A)(B)(C)(D)(E)
4. (A)(B)(C)(D)(E)   9. (A)(B)(C)(D)(E)
5. (A)(B)(C)(D)(E)   10. (A)(B)(C)(D)(E)

1. X buses can hold B children. There are Y children in D classes. How many buses are needed for each class?

(A) BD/XY
(B) BX/DY
(C) XY/BD
(D) DY/BX
(E) DX/BY

2. X ounces of gold cost Y dollars. Each ounce of silver costs $10. How many ounces of gold can be exchanged for D ounces of silver at the current rates?

(A) XYD/10
(B) 10/XYD
(C) 10Y/XD
(D) 10XD/Y
(E) Y/10XD

3. R pounds of potatoes can feed N people. How many pounds of potatoes must be served to feed P people?

(A) P/R
(B) NR/P
(C) N/PR
(D) PN/R
(E) PR/N

---

4. A hybrid car can travel 100 miles on 2 gallons of gas. If gas costs $3 per gallon, how many miles can the car travel on $24?

(A) 200
(B) 400
(C) 600
(D) 720
(E) 800

---

5. A car uses Y gallons of gas for every X miles traveled. Each gallon costs $5. If the car is traveling at 120 mph, what is the dollar cost of the gas used for each minute the car travels at the rates listed?

(A) 10Y/X
(B) 10X/Y
(C) X/10Y
(D) Y/10X
(E) 600Y/X

GO ON TO NEXT PAGE ⟩

CHAPTER 5: CHALLENGE QUESTIONS

Student-Produced Responses

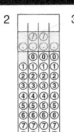

5.1

1. If cell phone company A charges $1.00 for the first minute, $0.05 a minute for the next 10 minutes, and $0.03 after that, and another cell phone company, B, charges $2.00 for the first minute, $0.04 per minute for the next 5 minutes, and $0.02 per minute after that, what would be the length of a phone call, in minutes, that would cost the same through either company?

2. A certain printing company charges $10 per book for the first 10 books printed. After that, the company charges $1 less per book for the next 5 books, $1 less than that for the following 5 books, and so on. How much would it cost, in dollars, to print 47 books?

5.2

3. Joey travels from home to work at X mph, and from work to home at 2X mph. If his entire round trip is 1.5 hours and the total distance traveled is represented by the expression X + 36, at what speed, in mph, does Joey travel from work to home?

5.3

4. If a chef can make a sandwich every 2 minutes and a professional eating contestant can eat a sandwich every 3 minutes, how much time, in minutes, will pass before there are 10 sandwiches on the table that remain to be eaten if the chef and the contestant start simultaneously?

5. If a freshman can solve a complex math problem in 5 minutes, a sophomore can solve one in 4 minutes, a junior can solve one in 3 minutes, and a senior can solve one in 2 minutes, how long will it take a four-student team with one student from each grade level to finish a 100-question exam?

5.4

6. If X beads can make C bracelets, and a machine can make G beads in H minutes, how many bracelets can be made in 4 hours?

Answer Here: _____

GO ON TO NEXT PAGE ⟩

Multiple-Choice	Student-Produced Responses

CHAPTER

5

REVIEW

1. ⒶⒷⒸⒹⒺ
2. ⒶⒷⒸⒹⒺ
3. ⒶⒷⒸⒹⒺ
4. ⒶⒷⒸⒹⒺ
ⒶⒷⒸⒹⒺ
ⒶⒷⒸⒹⒺ
ⒶⒷⒸⒹⒺ
ⒶⒷⒸⒹⒺ
ⒶⒷⒸⒹⒺ
ⒶⒷⒸⒹⒺ

1. G ounces of gold cost D dollars. S ounces of silver cost C dollars. How many ounces of gold are equivalent to 20 ounces of silver?

   (A) 20GC/DS
   (B) 20DS/GC
   (C) 20GS/DC
   (D) GC/20DS
   (E) DS/20GC

2. At the airport, every passenger pays $5 for the first piece of luggage that they stow and then pays $3 per piece after that. How much does it cost a passenger, in dollars, to stow 8 suitcases at this airport?

   (A) $12
   (B) $24
   (C) $26
   (D) $30
   (E) $36

3. A peregrine falcon flies for ten miles at 60 mph and then for two miles at 80 mph. What is the average speed for the falcon's flight in mph?

   (A) 58.48 mph
   (B) 62.61 mph
   (C) 63.48 mph
   (D) 63.61 mph
   (E) 68.01 mph

4. X bags of flour are used to bake Y cakes. 1 cake is enough to feed 4 people. How many bags of flour are needed to bake enough cakes to feed 50 people?

   (A) X/200Y
   (B) 200X/Y
   (C) 25X/2Y
   (D) 25Y/2X
   (E) 2Y/25X

5. Referring back to question #2, if the total cost of a person's luggage is $32, how many bags is this person trying to stow?

6. A cheetah, the world's fastest land animal, runs with sudden bursts of speed, but tires quickly. This animal can sustain a peak speed of 70 mph for up to ten minutes before it fatigues and drops its speed to 50 mph for another ten minutes, then can run at only 40 mph if it pushes on. If a cheetah runs for 40 minutes, what is its average speed for the entire 40-minute run?

7. Two cross country runners begin at the same starting line, but take different paths to the finish line. Runner 1 runs at an average speed of 8 mph and Runner 2 at an average speed of 10 mph. However, Runner 2 takes a path that is ten miles longer than the path taken by Runner 1. The two runners reach the finish line together, resulting in a tie. How many miles did Runner 2 run?

8. Shawn can clear the snow from a driveway in 40 minutes. Jonathan can clear the same driveway in one hour. How many minutes would it take them working together to clear the driveway of snow?

GO ON TO NEXT PAGE ▷

Multiple-Choice

Student-Produced Responses

CHAPTER 1 - 5
CUMULATIVE REVIEW

1 Ⓐ Ⓑ Ⓒ Ⓓ Ⓔ
2 Ⓐ Ⓑ Ⓒ Ⓓ Ⓔ
3 Ⓐ Ⓑ Ⓒ Ⓓ Ⓔ
4 Ⓐ Ⓑ Ⓒ Ⓓ Ⓔ
Ⓐ Ⓑ Ⓒ Ⓓ Ⓔ
Ⓐ Ⓑ Ⓒ Ⓓ Ⓔ
Ⓐ Ⓑ Ⓒ Ⓓ Ⓔ
Ⓐ Ⓑ Ⓒ Ⓓ Ⓔ
Ⓐ Ⓑ Ⓒ Ⓓ Ⓔ
Ⓐ Ⓑ Ⓒ Ⓓ Ⓔ

1. In an ant colony, there are 5 worker ants for every 3 warrior ants. If there are 200 worker ants, how many warrior ants are there?

(A) 60
(B) 120
(C) 180
(D) 240
(E) 320

2. In a storewide sale, all items are 20% off. If an item usually costs x dollars, how much will it cost during the sale in terms of x?

(A) .02x
(B) .08x
(C) .20x
(D) .80x
(E) 1.20x

3. The average of 4 numbers is 20. When one of the numbers is removed, the average drops to 15. What was the number that was removed?

(A) 35
(B) 40
(C) 45
(D) 50
(E) 55

4. If a certain number is divided by 20, it leaves a remainder of 3. If that number is doubled and then divided by 5, what is the remainder?

(A) 1
(B) 2
(C) 3
(D) 4
(E) 5

5. On a scenic drive, Tom drives from his home to a lookout point at an average speed of 20 mph and returns home along the same route at an average speed of 40 mph. If it takes a total of 3 hours for Tom to make the entire round-trip, how many hours does it take for him to reach the lookout?

6. In a library there are 3 times as many adult fiction books as adult non-fiction books, and the number of adult books (fiction and non-fiction together) is twice the number of children's books. If there are 1800 books in this library, how many adult fiction books are there?

7. The height of a tree increases by 6 feet every year. If the height of the tree is 30 feet in 2002, how many feet tall will the tree be in 2014?

8. In a mine, the ratio of pink diamonds to white diamonds is 4:5. There are currently 3600 diamonds that have been unearthed from this mine. How many more white diamonds need to be dug up in order to make the ratio of pink to white 1:2?

9. A skin-care company sells a 3-part set of beauty items. The 1st purchase in this set is considered a trial and is offered at the lowered rate of $25. Every purchase after the trial purchase is considered a refill and is sold at the full retail price of $45. Kesha has bought 5 sets of this 3-part system. How many dollars has she paid?

STOP

Chapter 6: The XY-Coordinate System

5 Sections
62 Practice Questions

6.1 Slopes

6.1.1 All About Slopes and Lines

Slope is a measure of how steep a line is. There are different ways to interpret and work with slopes, so make sure you are familiar with all of the equivalent expressions for the slope:

$$\text{Slope} = \frac{\text{Rise}}{\text{Run}} = \frac{\text{Change in } y}{\text{Change in } x} = \frac{y_2 - y_1}{x_2 - x_1} \quad \text{...from the two points on a line } (x_1, y_1) \text{ and } (x_2, y_2).$$

The equation of a line can be written in the following forms:

General Equations	Slope	Y-Intercept	X-Intercept
$y = mx + b$	m	b	Set y=0 , Solve for x.
$y - y_1 = m(x - x_1)$	m	$y_1 - mx_1$	Set y=0 , Solve for x.
$Ax + By = C$	-A/B	C/B	C/A

Further Notes on Lines and Slopes:

Line Type	Slopes
Parallel	Same Slope (Systems with Parallel Lines have no solution.)
Perpendicular	Opposite Reciprocal Slope (Systems with Perp. Lines have one solution.)
Positive Slope (m > 1)	Rises from Left to Right, Steep
Positive Slope (m = 1)	Rises from Left to Right at 45^0
Positive Slope (0 < m < 1)	Rises from Left to Right Shallow
Negative Slope (m < -1)	Falls from Left to Right, Steep
Negative Slope (m = -1)	Falls from Left to Right at 45^0
Negative Slope (-1 < m < 0)	Falls from Left to Right Shallow
Horizontal Line	Slope = 0 (Equation is **y = #**)
Vertical Line	Slope is infinite (Equation is **x = #**)

SAT Example and Technique Application:

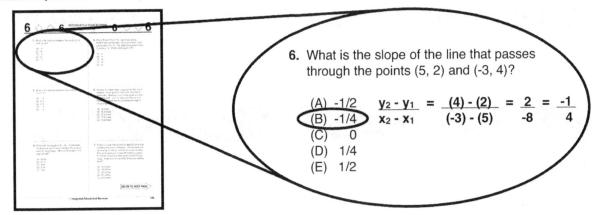

6. What is the slope of the line that passes through the points (5, 2) and (-3, 4)?

(A) -1/2
(B) -1/4
(C)   0
(D) 1/4
(E) 1/2

$$\frac{y_2 - y_1}{x_2 - x_1} = \frac{(4) - (2)}{(-3) - (5)} = \frac{2}{-8} = \frac{-1}{4}$$

6.1.1 **Let's Practice:**

1. Ⓐ Ⓑ Ⓒ Ⓓ Ⓔ   6. Ⓐ Ⓑ Ⓒ Ⓓ Ⓔ
2. Ⓐ Ⓑ Ⓒ Ⓓ Ⓔ   7. Ⓐ Ⓑ Ⓒ Ⓓ Ⓔ
3. Ⓐ Ⓑ Ⓒ Ⓓ Ⓔ   8. Ⓐ Ⓑ Ⓒ Ⓓ Ⓔ
4. Ⓐ Ⓑ Ⓒ Ⓓ Ⓔ   9. Ⓐ Ⓑ Ⓒ Ⓓ Ⓔ
5. Ⓐ Ⓑ Ⓒ Ⓓ Ⓔ   10. Ⓐ Ⓑ Ⓒ Ⓓ Ⓔ

1. What is the equation of the line that passes through the points (-6, 2) and (8, -4)?

(A) y = - 3/7x - 4/7
(B) y = - 3/7x - 8/7
(C) y =   5/8x - 3/8
(D) y =   3/8x + 3/8
(E) y =   3/8x + 5/8

2. What is the slope of the line 3x + 6y = 4?

(A)   2
(B)   -2
(C)   1/2
(D) -1/2
(E)   2/3

3. In the xy-plane, line *m* has the equation 4x + 2y = 5 and is perpendicular to line *n*. What is the slope of line *n*?

(A)   2
(B)   -2
(C)   1/2
(D) -1/2
(E)   5/2

4. What is the equation of the line that passes through the point (2, 5) and is perpendicular to the line with the equation y = -3x + 4?

(A) y = -3x + 1
(B) y = -3x + 11
(C) y = 1/3x + 13/3
(D) y = 1/3x - 13/3
(E) y = -1/3x + 13/3

5. Line *l* is parallel to line *k*, which has an equation of 5y = 2x + 20. What is the y-intercept of line *l* if this line passes through point (1, -3)?

(A) -17/5
(B) -13/5
(C)   -7/5
(D)   4/5
(E) 17/5

6. Line *l* has the equation 3x + *b*y = 4, where *b* is a constant. The line 3x + 7y = 6 is parallel to line *l*. What is the y-intercept of line *l*?

(A) 1/7
(B) 2/7
(C) 3/7
(D) 4/7
(E) 4/5

GO ON TO NEXT PAGE ⟩

6.1.2 **Intersecting Lines**

If two lines intersect, the point of their intersection will satisfy the linear equations of each line. This point of intersection can be found by setting the linear equations equal to each other, or through the elimination method.

SAT Example and Technique Application:

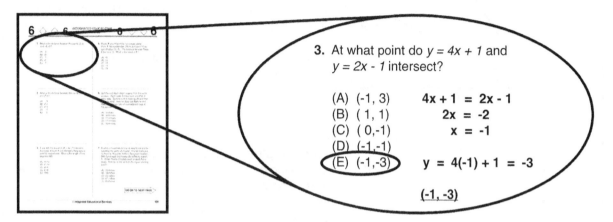

3. At what point do $y = 4x + 1$ and $y = 2x - 1$ intersect?

(A) (-1, 3) $4x + 1 = 2x - 1$
(B) (1, 1) $2x = -2$
(C) (0, -1) $x = -1$
(D) (-1, -1)
(E) (-1, -3) $y = 4(-1) + 1 = -3$

$(-1, -3)$

6.1.2 **Let's Practice:**

1. Ⓐ Ⓑ Ⓒ Ⓓ Ⓔ   6. Ⓐ Ⓑ Ⓒ Ⓓ Ⓔ
2. Ⓐ Ⓑ Ⓒ Ⓓ Ⓔ   7. Ⓐ Ⓑ Ⓒ Ⓓ Ⓔ
3. Ⓐ Ⓑ Ⓒ Ⓓ Ⓔ   8. Ⓐ Ⓑ Ⓒ Ⓓ Ⓔ
4. Ⓐ Ⓑ Ⓒ Ⓓ Ⓔ   9. Ⓐ Ⓑ Ⓒ Ⓓ Ⓔ
5. Ⓐ Ⓑ Ⓒ Ⓓ Ⓔ   10. Ⓐ Ⓑ Ⓒ Ⓓ Ⓔ

1. If line m is $y = 3x - 4$ and line n is $y = 4x - 3$, what is their point of intersection?

(A) (-1, 7)
(B) (1, 1)
(C) (0, 1)
(D) (-1, -1)
(E) (-1, -7)

2. If the equation for line k is $4y - x = -20$ and the equation for line l is $4y - 3x = 4$, at what x-value do the lines intersect?

(A) -12
(B) -8
(C) -4
(D) 6
(E) 12

3. Line f is parallel to the line $y = 2x - 7$ and passes through the point (3, 4). Line g is a line perpendicular to the line $y = 2x - 7$ and passes through the point (4, 3). What is the point of intersection of lines f and g ?

(A) (11/5, 18/5)
(B) (14/5, 18/5)
(C) (17/5, 11/5)
(D) (2/3, 3/4)
(E) (1/5, 3/4)

4. The equation for line k is $y = 1/5x + 4$ and line l is $y = 1/7x + b$. If lines k and l intersect at (a, 9/4), what is the value of b?

(A) 3/2
(B) 5/2
(C) 7/2
(D) 9/2
(E) 11/2

GO ON TO NEXT PAGE ⟩

6.1.3 Applying Slopes and Lines

Practice Examples

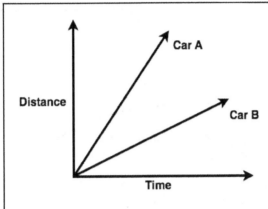

Explained conceptually, slope is the rate of change of y over x. In the graph to the left, the slopes of the lines represent the rate at which <u>distance</u> (y) changes over <u>time</u> (x) which means that these slopes represent the respective speeds of the two cars. The steeper the line, the greater the slope.

Which car is traveling at a faster rate?
<u>Car A</u>, since it has a steeper slope than Car B.

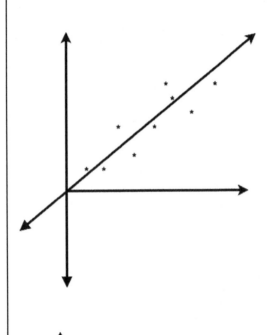

A **<u>Scatterplot</u>** is a cluster of data points taken from some kind of experiment or study and indicates a trend or relationship involving y and x. In the figure to the left, the data points appear to progress along a broken line.

Since you cannot connect all the points with a single line that would relate the two variables, the next best thing is to draw a **<u>best fit line</u>**, which is a line that divides the cluster of points approximately in half. For example, the line to the left seems to cut a middle path through the points. This is good enough!

Which equation below appears to be the best fit line in the picture?

a) y = 3x + 2
b) y = x
c) y = 1/2x

<u>B</u> would be the best answer, since the line appears to go through (0,0) and appears close enough to 45⁰.

More Practice:

Three specimens of the same plant, A, B, and C, are given varying amounts of water and light to determine the optimal conditions for the growth of this plant. Which plant received the highest ratio of water to light?

The correct answer is <u>A</u>.

6.2 Distance Formula

Any distance problem involving two points in the xy-plane can be solved by creating a right triangle and using the Pythagorean Theorem: $a^2 + b^2 = c^2$. The length of the hypotenuse, c, is the distance between the points.

Demonstration Example

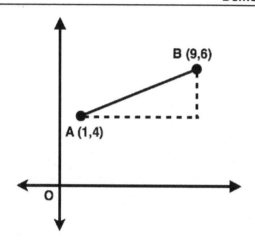

To find the distance from point A to point B, we will first find the absolute difference in x coordinates as one leg of a right triangle, and then the absolute difference in y coordinates as another leg of the same right triangle.

x: $|9 - 1| = 8$ y: $|6 - 4| = 2$

Then, using the pythagorean theorem, we can find the length of the hypotenuse:

$\sqrt{8^2 + 2^2} = \sqrt{68} \approx 8.25$

This is the same distance formula that you might have learned...

$$\text{Distance} = \sqrt{(x_2 - x_1)^2 + (y_2 - y_1)^2}$$

An easier way to remember this is with the following equation...

$$\text{Distance}^2 = (\Delta x)^2 + (\Delta y)^2$$

SAT Example and Technique Application:

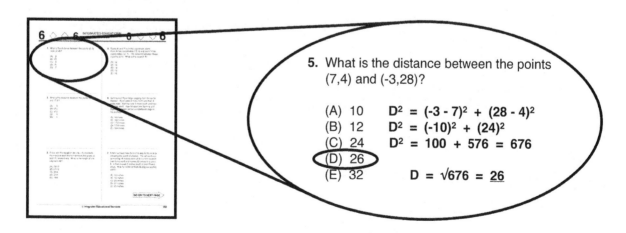

5. What is the distance between the points (7,4) and (-3,28)?

(A) 10 $D^2 = (-3 - 7)^2 + (28 - 4)^2$
(B) 12 $D^2 = (-10)^2 + (24)^2$
(C) 24 $D^2 = 100 + 576 = 676$
(D) 26
(E) 32 $D = \sqrt{676} = \underline{26}$

6.2 **Let's Practice:**

1. Ⓐ Ⓑ Ⓒ Ⓓ Ⓔ   6. Ⓐ Ⓑ Ⓒ Ⓓ Ⓔ
2. Ⓐ Ⓑ Ⓒ Ⓓ Ⓔ   7. Ⓐ Ⓑ Ⓒ Ⓓ Ⓔ
3. Ⓐ Ⓑ Ⓒ Ⓓ Ⓔ   8. Ⓐ Ⓑ Ⓒ Ⓓ Ⓔ
4. Ⓐ Ⓑ Ⓒ Ⓓ Ⓔ   9. Ⓐ Ⓑ Ⓒ Ⓓ Ⓔ
5. Ⓐ Ⓑ Ⓒ Ⓓ Ⓔ   10. Ⓐ Ⓑ Ⓒ Ⓓ Ⓔ

1. What is the distance between the points (2,-5) and (-2,-2)?

(A) 3
(B) $\sqrt{3}$
(C) 5
(D) $\sqrt{5}$
(E) 7

2. What is the distance between the points (8, 3) and (7,0)?

(A) 3
(B) $\sqrt{10}$
(C) $\sqrt{13}$
(D) 4
(E) 5

3. A line with the equation $3x + 4y = 5$ intersects the x-axis at point A and intersects the y-axis at point B, respectively. What is the length of line segment AB?

(A) 25/12
(B) 21/12
(C) 25/6
(D) 21/6
(E) 19/6

4. Jun rides his bike north at 60 mph. Kim rides her bike east at 80 mph. Assuming that they both start at the same location, how far apart are Jun and Kim after one hour of riding?

(A) 100 miles
(B) 120 miles
(C) $\sqrt{500}$ miles
(D) $\sqrt{1000}$ miles
(E) 10,000 miles

5. Points X and Y lie in the coordinate plane. Point X has coordinates (18, x) and point Y has coordinates (12, 7). The distance between these 2 points is 10. What is the value of X?

(A) 12
(B) 13
(C) 14
(D) 15
(E) 16

6. Sammy and Ryan begin jogging from the same location. Ryan runs 3 miles north and then 4 miles east. Sammy runs 5 miles south and then 12 miles west. How far apart are Sammy and Ryan after each runner completes both legs of his journey?

(A) 16 miles
(B) 16.5 miles
(C) 17.2 miles
(D) 17.9 miles
(E) 18.4 miles

7. A rat in a maze tries to find his way to the end by following the scent of cheese. The rat starts out by moving 18 inches north of its current location, then turns west and moves 20 inches to a point, A. It then moves 3 inches south of point A and stops. How far is the rat from its original starting point?

(A) 16 inches
(B) 18 inches
(C) 20 inches
(D) 21 inches
(E) 25 inches

GO ON TO NEXT PAGE ▷

6.3 Finding Midpoints

The <u>midpoint</u> of a segment is the *average* of the endpoints' coordinates. In order to find the coordinates of the midpoint, find the average of the x coordinates and the average of the y coordinates.

SAT Example and Technique Application:

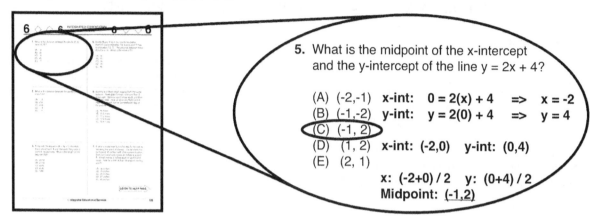

5. What is the midpoint of the x-intercept and the y-intercept of the line y = 2x + 4?

(A) (-2,-1)   x-int: 0 = 2(x) + 4   =>   x = -2
(B) (-1,-2)   y-int: y = 2(0) + 4   =>   y = 4
(C) (-1, 2)
(D) (1, 2)   x-int: (-2,0)   y-int: (0,4)
(E) (2, 1)

x: (-2+0) / 2   y: (0+4) / 2
Midpoint: (-1,2)

6.3 Let's Practice:

1. Ⓐ Ⓑ Ⓒ Ⓓ Ⓔ   6. Ⓐ Ⓑ Ⓒ Ⓓ Ⓔ
2. Ⓐ Ⓑ Ⓒ Ⓓ Ⓔ   7. Ⓐ Ⓑ Ⓒ Ⓓ Ⓔ
3. Ⓐ Ⓑ Ⓒ Ⓓ Ⓔ   8. Ⓐ Ⓑ Ⓒ Ⓓ Ⓔ
4. Ⓐ Ⓑ Ⓒ Ⓓ Ⓔ   9. Ⓐ Ⓑ Ⓒ Ⓓ Ⓔ
5. Ⓐ Ⓑ Ⓒ Ⓓ Ⓔ   10. Ⓐ Ⓑ Ⓒ Ⓓ Ⓔ

1. What is the midpoint of the segment defined by the endpoints (0,6) and (20,14)?

(A) (0,14)
(B) (20,6)
(C) (20,14)
(D) (10,10)
(E) (20,20)

2. What is the midpoint of the segment defined by the endpoints (14,-2) and (2,8)?

(A) (14,8)
(B) (2,-2)
(C) (6,3)
(D) (8,3)
(E) (8,5)

3. What is the midpoint of the x-intercept and the y-intercept of the line y = -4x + 6?

(A) (3/4,3)
(B) (-3/4,3)
(C) (3/2,3)
(D) (2/3,3)
(E) (-2/3,3)

4. Suppose that f(x) is a straight line, such that f(6) = 9, f(12) = 1, and f(9) = p. What is the value of p?

(A) 3
(B) 4
(C) 5
(D) 7
(E) 9

GO ON TO NEXT PAGE ▷

6.4 Coordinate Geometry

SAT problems will ask you to perform distance and midpoint calculations using coordinates of geometric shapes. These calculations involve area, perimeter, circumference, and other characteristics, asking you to synthesize knowledge of multiple subjects at once.

HINT: ALWAYS DRAW A PICTURE IF YOU ARE NOT GIVEN ONE.

6.4 Let's Practice:

1. Ⓐ Ⓑ Ⓒ Ⓓ Ⓔ   6. Ⓐ Ⓑ Ⓒ Ⓓ Ⓔ
2. Ⓐ Ⓑ Ⓒ Ⓓ Ⓔ   7. Ⓐ Ⓑ Ⓒ Ⓓ Ⓔ
3. Ⓐ Ⓑ Ⓒ Ⓓ Ⓔ   8. Ⓐ Ⓑ Ⓒ Ⓓ Ⓔ
4. Ⓐ Ⓑ Ⓒ Ⓓ Ⓔ   9. Ⓐ Ⓑ Ⓒ Ⓓ Ⓔ
5. Ⓐ Ⓑ Ⓒ Ⓓ Ⓔ   10. Ⓐ Ⓑ Ⓒ Ⓓ Ⓔ

1. A circle lies in the coordinate plane. This circle has its center at (6,8) and a radius of ten. How many total points of intersection does the circle have with the set of axes?

(A) 0
(B) 1
(C) 2
(D) 3
(E) More than 3

2. What is the distance between points B and C in the drawing below?

(A $\sqrt{2}$
(B) 2
(C) $2\sqrt{2}$
(D) $2\sqrt{3}$
(E) $4\sqrt{2}$

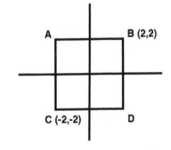

3. Triangle ABC lies in the coordinate plane. Point A has the coordinates (2,7) and point C has the coordinates (8,9). If D is the midpoint of AC, find the distance from B to D if B's coordinates are (4,12)?

(A) $\sqrt{11}$
(B) $\sqrt{12}$
(C) $\sqrt{17}$
(D) $\sqrt{19}$
(E) $\sqrt{21}$

4. A line with slope 3/2 passes through the origin and runs diagonally through 2 of the opposite corners of a rectangle. If the width of the rectangle is 8, what is the length of the rectangle?

(A) 12
(B) 10
(C) 8
(D) 6
(E) 4

GO ON TO NEXT PAGE ▷

6.5 Transformations

6.5.1 Reflection

Reflection is a process in which an object (usually a line, but often a point) is reproduced on the opposite side of a line (usually an axis).

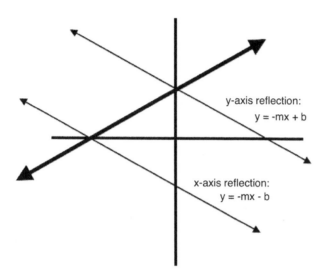

x-axis reflection: Both the <u>slope</u> and the <u>y-intercept</u> are opposite.

y-axis reflection: Only the <u>slope</u> is opposite.

y-axis reflection:
$y = -mx + b$

x-axis reflection:
$y = -mx - b$

Reflecting over the line y = x: Switch the x and the y and solve for y.

If a point or line is reflected across something other than an axis or the line x = y, use the coordinate system to count how many spaces the point or line "moves."

Practice Examples

1. **Find the equation of the following reflected lines:**

	y-axis	x-axis	y = x
y = 2x + 5			
y = -1/3x + 6			
y = 3/2x - 7			
y = 6x - 3			
y = 3			

2. **What are the x-intercept and the y-intercept of the line resulting from a reflection of the line** *y = x + 6* **across the line** *y = 6*?

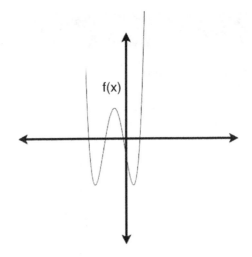

3. **Which of the following is the resulting graph when the graph of f(x) is reflected across the line y = x ?**

A.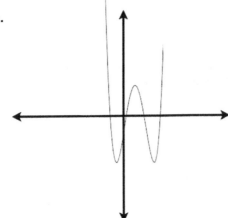

B.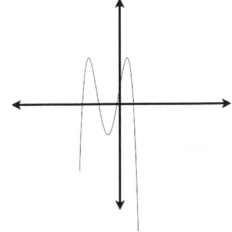

C.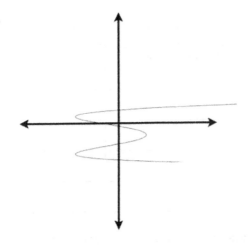

D.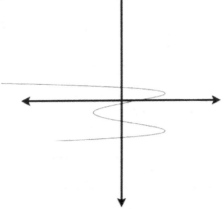

6.5.2 Rotation

Rotation about a point (usually the origin) occurs when a figure is "turned" a certain number of degrees as if it were pinned down to that point. On the SAT, most rotation problems rely on visual perceptions.

Practice Problem:
Which of the following results when the triangle below is rotated 90⁰ clockwise about the origin?

Pre-Image:

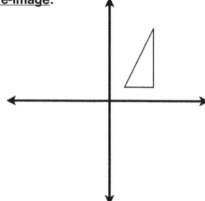

A.

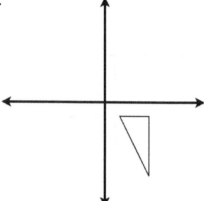

B.

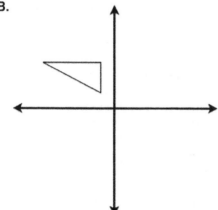

C.

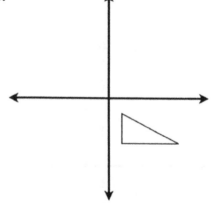

With rotation, visually move one side of the figure in a turning or revolving motion as if it were pinned or tied to the rotation point. Here, if you "tie" the bottom side of the original triangle to the origin, and turn it 90⁰, you get C. Most of the time, it is easiest to take your pencil and place it on the origin of the coordinate plane. Then, you can literally rotate the test clockwise or counter-clockwise by the appropriate amount. At this point, you can sketch the image on a piece of scrap paper, then you can look for the appropriate answer.

6.5.3 **Translation**

Translation is a transformation where the shape maintains its orientation (slope for a line, relative position of the vertices of a polygon) but shifts a certain distance along the plane. All translations can be understood as the combination of a horizontal x-shift and a vertical y-shift.

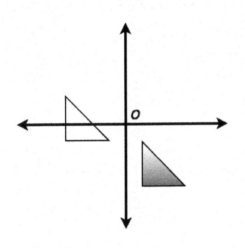

Practice Problem:

In the figure to the left, the unshaded triangle has undergone a translation to become the shaded triangle. Which of the following best describes this translation?

A. **A positive y-translation and a larger positive x-translation.**

B. **A negative y-translation and a larger positive x-translation.**

C. **A negative y-translation and a smaller positive x-translation.**

D. **A positive y-translation and a smaller positive x-translation.**

E. **A positive y-translation and a larger negative x-translation.**

6.5.4 **Transformation Practice**

Draw the results of the following transformations of the given triangle.

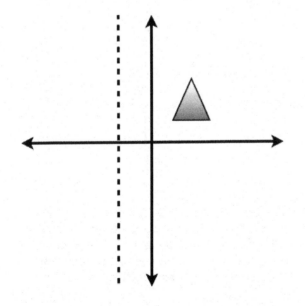

1. **Rotation of 180⁰ about the origin.**
2. **Reflection across the y-axis**
3. **Rotation of 360⁰ about the origin**
4. **Reflection across the x-axis**
5. **Reflection across the dotted line**
6. **Positive x-shift translation**
7. **Negative y-shift translation**
8. **Reflection across the line y = x**
9. **Rotation of 180⁰ about the origin, followed by a reflection across the y-axis**
10. **Which combinations of the above-mentioned transformations could produce the same end result?**

    I.   **3 and 8**
    II. **5 and 2 followed by 6**
    III. **4 and 9**

6 ◇ ◇ 6   6 ◇ ◇ 6

Unauthorized copying or
reuse of any part of this
page is illegal.

CHAPTER 6: CHALLENGE QUESTIONS

Student-Produced Responses

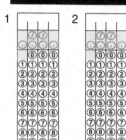 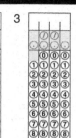    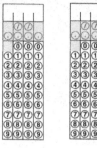 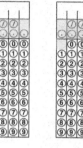 

6.1

1. Line *m* is tangent to a circle at the point (4,1). If the circle is centered at the origin, what is the y value of the y-intercept of line *m*?

2. Two lines intersect at the point (3,2). The equation of one line is y = mx + b and the equation of the other line is y = bx + m. What is the value of m?

6.2

3. A parallelogram has a height of 5 and a base length of 20. If the base is parallel to the x-axis, and the angled sides have a slope of 5/12, what is the distance between the two farthest points of the parallelogram?

6.3

4. A is the vertex of the function f(x) = 2x² - 5x - 3. B is the negative root of the same function. What is the x coordinate of the midpoint of line segment AB?

6.4

5. An equilateral triangle is centered at the origin. If one of the vertices is at (4√2, 4), what is the length of one side of the triangle?

6.5

6. If a line with the equation 1/2x + 2/3y = 5 is reflected over the x-axis, then over the line y = x, what is the slope of the new line?

7. Suppose that a circle is centered at the origin and that a line with a slope of -1/3 is tangent to the circle. If the circle has a radius of 5, what is the y-intercept of the tangent line?

GO ON TO NEXT PAGE ▷

Multiple-Choice

Student-Produced Responses

CHAPTER **6** REVIEW

1 Ⓐ Ⓑ Ⓒ Ⓓ Ⓔ
2 Ⓐ Ⓑ Ⓒ Ⓓ Ⓔ
3 Ⓐ Ⓑ Ⓒ Ⓓ Ⓔ
Ⓐ Ⓑ Ⓒ Ⓓ Ⓔ
Ⓐ Ⓑ Ⓒ Ⓓ Ⓔ
Ⓐ Ⓑ Ⓒ Ⓓ Ⓔ
Ⓐ Ⓑ Ⓒ Ⓓ Ⓔ
Ⓐ Ⓑ Ⓒ Ⓓ Ⓔ
Ⓐ Ⓑ Ⓒ Ⓓ Ⓔ
Ⓐ Ⓑ Ⓒ Ⓓ Ⓔ

1. The table below shows mass as a function of the volume of a certain liquid for every 10 cubic centimeters. Which of the following graphs shows the correct relationship between mass and volume?

Volume (cc)	Mass (g)
0	0
10	7.55
20	15.1
30	22.65

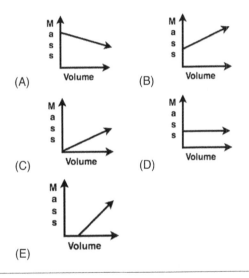

(A)   (B)

(C)   (D)

(E)

2. A line with a slope of 2 forms a 90 degree angle with a line that goes through the points (2,3) and (1,m). What is the value of m?

(A) 2
(B) 2.5
(C) 3
(D) 3.5
(E) 4

3. What is the value of Z in the following system of equations if there is no solution?

$$3x + 4y = 8 \quad \text{and} \quad 6x + Zy = 10$$

(A) 2
(B) 4
(C) 6
(D) 7
(E) 8

4. A line with equation $4x + 7y = B$ passes through the point (1,2). What is the value of B?

5. The following points are collinear: (1,3), (p,7), and (4,9). What is the value of p?

6. Points X, Y, W, and Z mark the four corners of a square. Points X (-3,-2) and W (3,3) are two opposite corners of the square. If V is the midpoint of Y and W, what is the area of triangle XVZ.

7. What is the slope of a line that is parallel to the line *m*, shown below, after line *m* has been reflected over the y-axis?

(0,12)

(5,0)

GO ON TO NEXT PAGE ▷

Multiple-Choice

Student-Produced Responses

CHAPTER
1 - 6
CUMULATIVE
REVIEW

1 Ⓐ Ⓑ Ⓒ Ⓓ Ⓔ
2 Ⓐ Ⓑ Ⓒ Ⓓ Ⓔ
3 Ⓐ Ⓑ Ⓒ Ⓓ Ⓔ
4 Ⓐ Ⓑ Ⓒ Ⓓ Ⓔ
Ⓐ Ⓑ Ⓒ Ⓓ Ⓔ
Ⓐ Ⓑ Ⓒ Ⓓ Ⓔ
Ⓐ Ⓑ Ⓒ Ⓓ Ⓔ
Ⓐ Ⓑ Ⓒ Ⓓ Ⓔ
Ⓐ Ⓑ Ⓒ Ⓓ Ⓔ
Ⓐ Ⓑ Ⓒ Ⓓ Ⓔ

1. A quilt is made using strips of alternating colors: red, blue, black, and indigo. If there are 178 total strips, what is the color of the last strip?

(A) red
(B) blue
(C) black
(D) indigo
(E) some other color

2. If one gram of protein contains 8 calories and one gram of fat contains 9 calories. What is the total amount of calories in 50 grams of protein and 100 grams of fat?

(A) 1000
(B) 1200
(C) 1300
(D) 1500
(E) 1600

3. A certain meal is 2 parts protein, 2 parts carbohydrate and 1 part fat. What is the total amount of fat present in a 100 ounce meal?

(A) 20
(B) 25
(C) 40
(D) 50
(E) 80

4. $K^X \cdot K^Y = K^3$ and $K^X/K^Y = K^{-5}$. What is X?

(A) -2
(B) -1
(C) 0
(D) 1
(E) 2

5. M percent of 50 is 3 percent of 25. What is M percent of 90?

6. A car that was initially bought for $40,000 decreases in value by 3% for every 500 miles its driven. What is the value of the car, in dollars, after it has traveled 50,000 miles? *(Round to the nearest dollar.)*

7. The scatterplot below shows the number of books read (y-axis) as compared to the number of video games played (x-axis) for 7 children in a certain month. For how many students was the number of video games played greater than the number of books read?

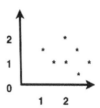

8. A store is offering a deal on certain items. If a customer buys 2 items at the regular price, he may then purchase additional units of that same item at half price. If Martin buys boxes of mint chocolate chip cookies valued at $3 each, how much does he pay if he purchases a total of 8 boxes?

S T O P

Chapter 7: Functions
7 Sections
51 Practice Questions

7.1 Quadratic Functions

There are a few terms you should be familiar with when working with parabolas, which are the graphs of quadratic equations. These equations always contain an x^2 term:

Vertex: The vertex of a parabola is the lowest point on the curve of a parabola when it opens upward and the highest point on the curve of a parabola when it opens downward.

Axis of Symmetry: The axis of symmetry is an imaginary line which divides the parabola into two symmetrical halves. It runs through the vertex and, therefore, **it equals the x-coordinate of the vertex.** The axis of symmetry is also the midpoint of the roots of a quadratic equation.

Quadratic Form	Vertex / Axis of Symmetry	Meaning of Constants
General Form: $ax^2 + bx + c$	Axis of Symmetry: $x = -b/2a$ Vertex: Use the equation above to find the x-coordinate, then plug that into the original equation to find the y-coordinate.	a -> If positive, the graph opens upward. If negative, the graph opens downward. If $|a| > 1$, then the graph will appear narrower than normal. If $0 < |a| < 1$, then the graph will appear wider than normal. c -> The y-intercept of the graph.
Vertex Form: $a(x - h)^2 + k$	Axis of Symmetry: $x = h$ Vertex: (h, k)	a -> The constant a in vertex form follows the same rules as in the general form above. h -> The h is the horizontal shift of the graph. k -> The k is the vertical shift of the graph.

SAT Example and Technique Application:

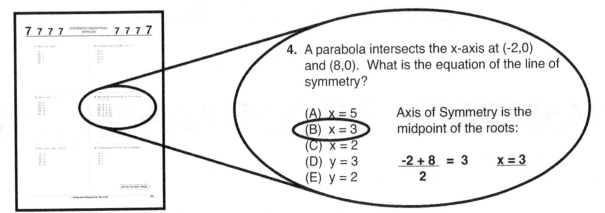

4. A parabola intersects the x-axis at (-2,0) and (8,0). What is the equation of the line of symmetry?

(A) x = 5
(B) x = 3
(C) x = 2
(D) y = 3
(E) y = 2

Axis of Symmetry is the midpoint of the roots:

$$\frac{-2 + 8}{2} = 3 \qquad x = 3$$

7.1 Let's Practice:

1. Ⓐ Ⓑ Ⓒ Ⓓ Ⓔ   6. Ⓐ Ⓑ Ⓒ Ⓓ Ⓔ
2. Ⓐ Ⓑ Ⓒ Ⓓ Ⓔ   7. Ⓐ Ⓑ Ⓒ Ⓓ Ⓔ
3. Ⓐ Ⓑ Ⓒ Ⓓ Ⓔ   8. Ⓐ Ⓑ Ⓒ Ⓓ Ⓔ
4. Ⓐ Ⓑ Ⓒ Ⓓ Ⓔ   9. Ⓐ Ⓑ Ⓒ Ⓓ Ⓔ
5. Ⓐ Ⓑ Ⓒ Ⓓ Ⓔ   10. Ⓐ Ⓑ Ⓒ Ⓓ Ⓔ

1. Which function of y is displayed in the graph below?

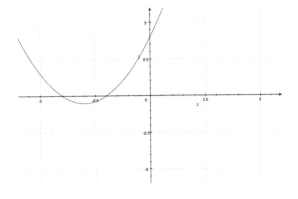

(A) $(1/2)x^2 + 3x + 4$
(B) $x^2 - 4x + 2$
(C) $-2x^2 - 6x - 2$
(D) $x^2 - 3x - 4$
(E) $-x^2 + 7x + 6$

2. A quadratic function has the equation:

$$f(x) = 2x^2 + 8x + 6$$

What is the slope of the line that goes through both the vertex of this parabola and the origin?

(A) 1/4
(B) 1/2
(C) 1
(D) -1/2
(E) -1/4

3. A parabola with the line of symmetry x = 1 has x-intercepts of (j, 0) and (5, 0). What is j?

(A) 3
(B) 2
(C) 0
(D) -2
(E) -3

GO ON TO NEXT PAGE ▷

116

7.2 Transforming Higher Order Functions

In general, a function f(x) can undergo transformations "inside" of the parentheses (causing x-axis translations) or "outside" of the parentheses (causing y-axis translations). This is similar to what we just saw with quadratic functions, mainly equations in vertex form.

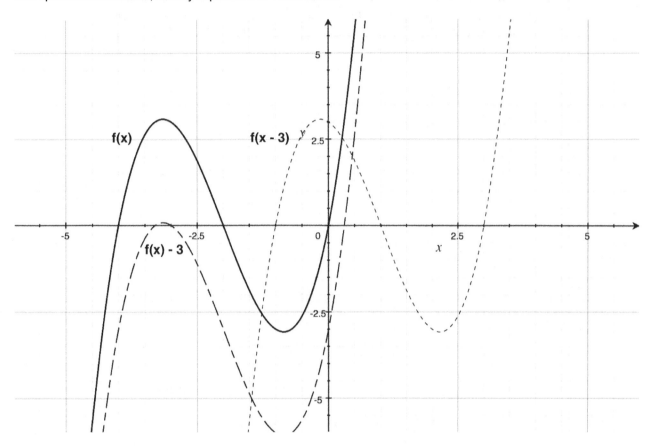

Demonstration Example:

Inside: f(x - 3) will cause a shift to the **right** of 3 units. Similarly, f(x + 3) will cause a shift to the **left** of 3 units. Just remember an inside change is a horizontal shift, opposite the direction of the sign.

Outside: f(x) - 3 will cause a shift **downward** of 3 units and f(x) + 3 will cause a shift **upward** of 3 units. An outside change is a vertical shift of the graph, exactly as stated.

Multiplier: Multiplying f(x) by a constant will not shift the graph, but it will stretch the graph vertically. Multiplying by a negative constant will flip the graph vertically over the y-axis. If |a| > 1, the graph will appear more narrow, and if 0 < |a| < 1, the graph will appear wider.

Practice:

Practice sketching the graphs of the following transformations of the original function, f(x), shown above.

1. f(x - 2) + 2   **2.** f(x + 3) - 1   **3.** - f(x)

4. 2f(x) + 1   **5.** f(x) + 5   **6.** f(x + 2)

7.3 Intersecting Graphs

When two graphs intersect at a point, the x and y values are equal.

Demonstration Example

Demo: Here you see two graphs: $f(x) = x^2 + 6x + 9$ and $g(x) = x^2 - 4x + 4$, which intersect at the point (x, y). What is the value of x + y?

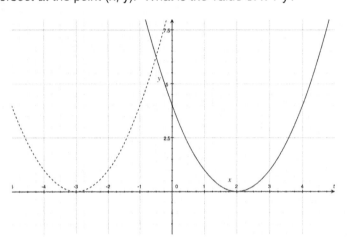

In order to solve the equation, we set f(x) and g(x) equal to each other and solve for x:

$$x^2 + 6x + 9 = x^2 - 4x + 4$$
$$10x = -5 \quad \rightarrow \quad x = -1/2$$

Then, we plug in our x-value in order to find y: $y = (-1/2)^2 - 4(-1/2) + 4$

$$(-1/2 , 25/4) \quad \rightarrow \quad x + y = -1/2 + 25/4 = \underline{23/4}.$$

SAT Example and Technique Application:

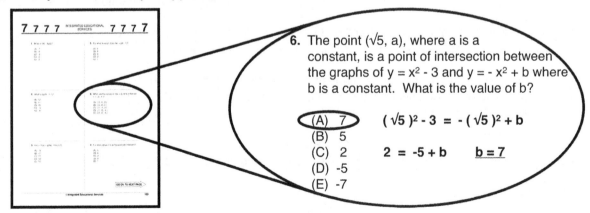

6. The point ($\sqrt{5}$, a), where a is a constant, is a point of intersection between the graphs of $y = x^2 - 3$ and $y = -x^2 + b$ where b is a constant. What is the value of b?

(A) 7
(B) 5
(C) 2
(D) -5
(E) -7

$(\sqrt{5})^2 - 3 = - (\sqrt{5})^2 + b$

$2 = -5 + b \qquad \underline{b = 7}$

7.3 Let's Practice:

1. Ⓐ Ⓑ Ⓒ Ⓓ Ⓔ   6. Ⓐ Ⓑ Ⓒ Ⓓ Ⓔ
2. Ⓐ Ⓑ Ⓒ Ⓓ Ⓔ   7. Ⓐ Ⓑ Ⓒ Ⓓ Ⓔ
3. Ⓐ Ⓑ Ⓒ Ⓓ Ⓔ   8. Ⓐ Ⓑ Ⓒ Ⓓ Ⓔ
4. Ⓐ Ⓑ Ⓒ Ⓓ Ⓔ   9. Ⓐ Ⓑ Ⓒ Ⓓ Ⓔ
5. Ⓐ Ⓑ Ⓒ Ⓓ Ⓔ   10. Ⓐ Ⓑ Ⓒ Ⓓ Ⓔ

1. The graph of $y = x^2 - 9$ intersects the graph of a line at $(0, a)$ and $(4, b)$. What is the greatest possible value of the slope of the line?

(A) 4
(B) 2
(C) 0
(D) -2
(E) -4

2. In the graph below, for how many values of x does $f(x) = 1/4$?

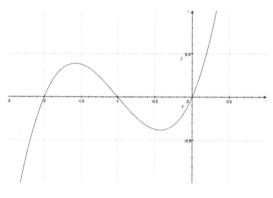

(A) 0
(B) 1
(C) 2
(D) 3
(E) 4

7.4 Manipulating Functions and Composites

Demonstration Example

Demo: Let the function $f(x)$ be defined by $f(x) = 2x - 3$. If $2f(k) = 18$, what is the value of $f(3k)$?

The key to ALL of these problems is to understand that $f(x)$ basically defines a mathematical operation. Whatever number "x" is inside the parentheses, $f(x)$ means to double it and subtract 3. So if $x = k$, just plug in k wherever you see x. $f(k) = 2k - 3$ and $2f(k)$ is just double that: $4k - 6$.

Next, since $2f(k) = 18$, just set the expression we found before for $2f(k)$ equal to 18. So $4k - 6 = 18$ and $k = 6$.

Now that we have k, we can go back and find $f(3k)$. Since $k = 6$, $3k$ equals 18. Now just plug 3k back into the original function $f(x) = 2x - 3$ -->
$f(18) = 2(18) - 3 = 33$. Done!

SAT Example and Technique Application:

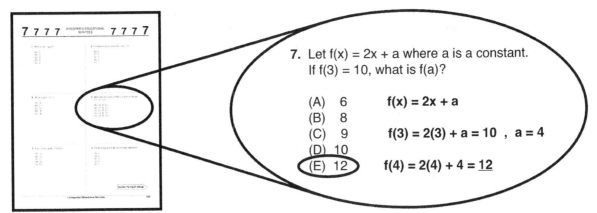

7. Let $f(x) = 2x + a$ where a is a constant. If $f(3) = 10$, what is $f(a)$?

(A)   6 $f(x) = 2x + a$
(B)   8
(C)   9 $f(3) = 2(3) + a = 10$, $a = 4$
(D)   10
(E)   12 $f(4) = 2(4) + 4 = \underline{12}$

7.4 Let's Practice:

1. Ⓐ Ⓑ Ⓒ Ⓓ Ⓔ   6. Ⓐ Ⓑ Ⓒ Ⓓ Ⓔ
2. Ⓐ Ⓑ Ⓒ Ⓓ Ⓔ   7. Ⓐ Ⓑ Ⓒ Ⓓ Ⓔ
3. Ⓐ Ⓑ Ⓒ Ⓓ Ⓔ   8. Ⓐ Ⓑ Ⓒ Ⓓ Ⓔ
4. Ⓐ Ⓑ Ⓒ Ⓓ Ⓔ   9. Ⓐ Ⓑ Ⓒ Ⓓ Ⓔ
5. Ⓐ Ⓑ Ⓒ Ⓓ Ⓔ   10. Ⓐ Ⓑ Ⓒ Ⓓ Ⓔ

1. Let $f(x) = x^2$. If $1/2\ f(b) = 32$, what is $f(2b)$?

(A)   4
(B)   16
(C)   64
(D)   256
(E) 1012

2. Let $f(x) = 2x^2 + 1$ and $g(x) = f(x) - 3$. If $g(k) = 96$, then what is $f(k)$?

(A) 99
(B) 98
(C) 49
(D) 48
(E) 24

3. Let $h(x) = \sqrt{(3x)}$. If $h(m) = 6$, what is the value of m?

(A)   9
(B) 12
(C) 15
(D) 18
(E) 36

4. Let $g(x) = x^2 + 6x + 9$ and let $h(x) = \sqrt{x}$. What is the value of $h(g(1))$?

(A)   4
(B)   16
(C)   64
(D)   256
(E) 1012

5. Let $f(x) = x^3 + 1$ and let $g(x) = \sqrt{x} + 1$. What is the value of $g(f(2))$?

(A) 1
(B) 2
(C) 4
(D) 8
(E) 12

GO ON TO NEXT PAGE ⟩

7.5 Symbol Questions

Demonstration Example

Demo: For all values of x, let x❖ be defined by x❖ = x^2 - 1. What equation is equivalent to (x❖)❖?

Conceptually, there is no difference between these peculiar-looking symbol questions and functions. Instead of using familiar notation such as f(x), these equations simply use a different symbol for f.

1. **Start with what is in the parentheses and perform the defined operation:**

$$x❖ = x^2 - 1$$

2. **Now deal with the second symbol that is outside the parenthesis. This is basically a composite function, similar to f(g(x)). You first complete the inner function, g(x), and then plug your answer into the outer function, f(x):**

$$(x❖)❖ = (x^2 - 1)❖ = (x^2 - 1)^2 - 1 = \underline{x^4 - 2x^2}$$

SAT Example and Technique Application:

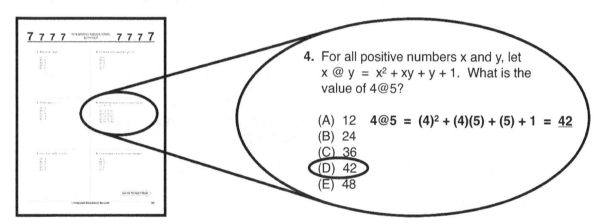

4. For all positive numbers x and y, let
   x @ y = x^2 + xy + y + 1. What is the value of 4@5?

   (A) 12   4@5 = $(4)^2$ + (4)(5) + (5) + 1 = **42**
   (B) 24
   (C) 36
   (D) 42
   (E) 48

7.5 Let's Practice:

1. Ⓐ Ⓑ Ⓒ Ⓓ Ⓔ   6. Ⓐ Ⓑ Ⓒ Ⓓ Ⓔ
2. Ⓐ Ⓑ Ⓒ Ⓓ Ⓔ   7. Ⓐ Ⓑ Ⓒ Ⓓ Ⓔ
3. Ⓐ Ⓑ Ⓒ Ⓓ Ⓔ   8. Ⓐ Ⓑ Ⓒ Ⓓ Ⓔ
4. Ⓐ Ⓑ Ⓒ Ⓓ Ⓔ   9. Ⓐ Ⓑ Ⓒ Ⓓ Ⓔ
5. Ⓐ Ⓑ Ⓒ Ⓓ Ⓔ   10. Ⓐ Ⓑ Ⓒ Ⓓ Ⓔ

1. For all positive integers z, let z# equal the product of all of the positive integers from 1 to z. What is the value of 8# - 5#?

   (A) 24,320
   (B) 36,000
   (C) 40,200
   (D) 40,320
   (E) 42,200

GO ON TO NEXT PAGE ⟩

2. For all numbers x and y, let $x \mathcal{M} y = x^2y + xy^2$. For all numbers a, b, and c, which of the following must be true?

      I. $a \mathcal{M} b = b \mathcal{M} a$
      II. $(a + b) \mathcal{M} c = (b + c) \mathcal{M} a$
      III. $b \mathcal{M} c = bc(c + b)$

(A) I only
(B) II only
(C) III only
(D) I and II
(E) I and III

3. For all real numbers x and y such that x does not equal y, let $x \, \Omega \, y = (x + y) / (x - y)$. If $a \, \Omega \, b = 125$ and $a + b = 5$, what is the product of a and b?

(A) 5.75
(B) 6.00
(C) 6.25
(D) 6.75
(E) 7.25

7.6 Plugging in Points to Find a Constant

Demonstration Example

Demo: Let the function $f(x) = ab^x$ where a and b are both constants. The following table contains corresponding values of x and f(x) for the function. Find the values of a and b.

x	0	1	2
f(x)	3	6	12

By plugging in the first pair of coordinates from the table (0,3) into the function you get:
$$3 = ab^{(0)}, \text{ so, } \underline{a = 3}.$$

Since you know what $\underline{a}$ is, you can plug in another point to find b. If you use (1,6):
$$6 = 3b^{(1)}, \text{ so, } \underline{b = 2}.$$

SAT Example and Technique Application:

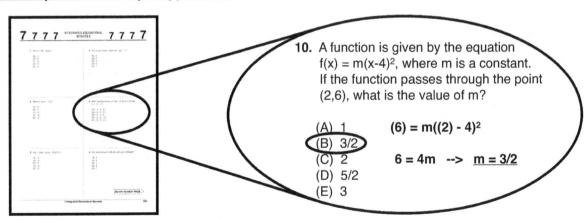

10. A function is given by the equation $f(x) = m(x-4)^2$, where m is a constant. If the function passes through the point (2,6), what is the value of m?

(A) 1 $(6) = m((2) - 4)^2$
(B) 3/2
(C) 2 $6 = 4m \; --> \; \underline{m = 3/2}$
(D) 5/2
(E) 3

7.6 **Let's Practice:**

1. Ⓐ Ⓑ Ⓒ Ⓓ Ⓔ   6. Ⓐ Ⓑ Ⓒ Ⓓ Ⓔ
2. Ⓐ Ⓑ Ⓒ Ⓓ Ⓔ   7. Ⓐ Ⓑ Ⓒ Ⓓ Ⓔ
3. Ⓐ Ⓑ Ⓒ Ⓓ Ⓔ   8. Ⓐ Ⓑ Ⓒ Ⓓ Ⓔ
4. Ⓐ Ⓑ Ⓒ Ⓓ Ⓔ   9. Ⓐ Ⓑ Ⓒ Ⓓ Ⓔ
5. Ⓐ Ⓑ Ⓒ Ⓓ Ⓔ   10. Ⓐ Ⓑ Ⓒ Ⓓ Ⓔ

1. The graph of $y = x^2 + kx - 8$ crosses the x axis at x=4. What is the value of k?

(A) 2
(B) 1
(C) 0
(D) -1
(E) -2

2. Let the function $f(x) = k(x-1)(x-4)$, where k is a constant. If $f(m+1) = 0$ where m is a constant, which of the following is a possible value for m?

(A) -3
(B) -2
(C) -1
(D) 0
(E) 1

3. In the table below, the function is linear and k and p are both constants. What is the value of k + p?

x	0	1	2
y	k	20	p

(A) 30
(B) 32
(C) 40
(D) 42
(E) 50

4. A square with an area of 100 is centered at the origin. A parabola with the equation $y = mx^2$ passes through the square's top two vertices. What is the value of m?

(A) 25
(B) 15
(C) 5
(D) 1
(E) 1/5

7.7 Interpreting Tables

x	f(x)	g(x)
-1	2	1
0	4	-1
1	3	0
2	6	4
3	4	2
4	5	5
5	3	-5

Demonstration Example

Demo: According to the the table above, what is 2f(3) + g(0)?

f(3) = 4 and g(0) = -1, so, 2(4) + (-1) = 7.

---

7 **7** **7** **7** **7** **7** **7** **7**

Unauthorized copying or reuse of any part of this page is illegal.

7.7 **Let's Practice:** **(Use the table on the previous page.)**

1. Ⓐ Ⓑ Ⓒ Ⓓ Ⓔ   6. Ⓐ Ⓑ Ⓒ Ⓓ Ⓔ
2. Ⓐ Ⓑ Ⓒ Ⓓ Ⓔ   7. Ⓐ Ⓑ Ⓒ Ⓓ Ⓔ
3. Ⓐ Ⓑ Ⓒ Ⓓ Ⓔ   8. Ⓐ Ⓑ Ⓒ Ⓓ Ⓔ
4. Ⓐ Ⓑ Ⓒ Ⓓ Ⓔ   9. Ⓐ Ⓑ Ⓒ Ⓓ Ⓔ
5. Ⓐ Ⓑ Ⓒ Ⓓ Ⓔ   10. Ⓐ Ⓑ Ⓒ Ⓓ Ⓔ

1. What is f(4) - 3g(2)?

(A) -7
(B) -5
(C) -3
(D) 0
(E) 2

2. What is 5g(3) - f(-1)?

(A) 12
(B) 11
(C) 10
(D) 9
(E) 8

3. h(x) = 3f(x) + g(3x). Find h(1).

(A) 9
(B) 10
(C) 11
(D) 12
(E) 13

4. For what value of x do f(x) and g(x) intersect?

(A) 5
(B) 4
(C) 3
(D) 2
(E) 1

5. For what x-value does f(x) = g(x - 1)?

(A) 5
(B) 4
(C) 3
(D) 2
(E) 1

6. What are the values of f(2x + 3) for x in the set { -1 , 0 , 1 }?

(A) { 3 , 4 , 3 }
(B) { 3 , 4 , 0 }
(C) { 2 , 4 , 3 }
(D) { 0 , 2 , -5 }
(E) { 0 , 4 , -5 }

GO ON TO NEXT PAGE

7 **7** **7** **7**   **7** **7** **7** **7**

Unauthorized copying or reuse of any part of this page is illegal.

CHAPTER 7: CHALLENGE QUESTIONS

Student-Produced Responses

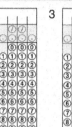

7.1

1. How many times does the function $f(x) = -1/2x^2 + 2x + 3$ intersect the line $y = 6$?

2. How many times does $y = x^2$ intersect $y = x^3$ on the interval $(0, 1]$?

7.2

3. Suppose that $f(x) = x^4 + 3x^2 + 4x + 2$ and that $g(x) = f(x - 3) - 2$. How many times does $f(x)$ intersect $g(x)$?

7.3

4. The line $y = 2x + 3$ intersects the function $y = x^2 + 2x - 3$ at points A and B. What is the distance from A to B? *(Round to the nearest whole number.)*

7.4

5. If $f(x) = 3x + b$, $g(x) = x - f(x)$, $h(x) = 2g(x)$, and $f(2) = -10$, what is the value of $h(b)$?

7.5

6. Suppose $x \star y = 3(x - y)^2$, $x \triangle y = (3xy)$, and $x \bigcirc y = x / 2y$. What is $((2\triangle3) \star (20 \bigcirc 5))$?

7.6

7. The table below shows a few points on a function. The function is quadratic, the vertex is located at the origin, and m and h are both constants. What is the value of $(h + 1) / m$?

x	f(x)
0	h
2	3/2
3	m

GO ON TO NEXT PAGE ⟩

---

7 7 7 7 7 **7 7 7**

Unauthorized copying or reuse of any part of this page is illegal.

Multiple-Choice

Student-Produced Responses

CHAPTER 7 REVIEW

1 Ⓐ Ⓑ Ⓒ Ⓓ Ⓔ
2 Ⓐ Ⓑ Ⓒ Ⓓ Ⓔ
3 Ⓐ Ⓑ Ⓒ Ⓓ Ⓔ
4 Ⓐ Ⓑ Ⓒ Ⓓ Ⓔ
Ⓐ Ⓑ Ⓒ Ⓓ Ⓔ
Ⓐ Ⓑ Ⓒ Ⓓ Ⓔ
Ⓐ Ⓑ Ⓒ Ⓓ Ⓔ
Ⓐ Ⓑ Ⓒ Ⓓ Ⓔ
Ⓐ Ⓑ Ⓒ Ⓓ Ⓔ
Ⓐ Ⓑ Ⓒ Ⓓ Ⓔ

5 6 7

1. Given the function $f(x) = x^2 - 4x + 3$, where A represents the y-intercept and B represents the function's lowest positive root, what is the distance from A to B?

(A) 2
(B) $2\sqrt{2}$
(C) 3
(D) $\sqrt{10}$
(E) $3\sqrt{2}$

2. The function $f(x) = x^2 + 6x + 5$ and the function $g(x) = x^2 - 4x - 5$ intersect at the point (a,b). What is the value of $b - a$?

(A) 0
(B) 1
(C) 2
(D) 3
(E) 5

3. The function $f(x)$ follows a linear model. If $f(4) = 7$, $f(10) = 9$, and $f(13) = b$, what is b?

(A) 15
(B) 14
(C) 12
(D) 11
(E) 10

4. The function $a \blacktriangleright b$ is defined by taking the number of positive factors of a and multiplying this number by b. What is the value of $24 \blacktriangleright 3$?

(A) 48
(B) 36
(C) 24
(D) 18
(E) 12

5. The functions $y = x^2$ and $y = b - x^2$ intersect each other in quadrant 1 at the point M. If the x-coordinate of M is 5, what is the value of b?

6. In the xy-coordinate plane, the graph of $x = y^2 - 4$ intersects the line k at $(12, a)$ and $(21, b)$. What is the greatest possible value for the slope of k?

7. A physics student charts the path of a ball thrown into the air. He finds that the vertical height, V(t), of the ball at time t can be modeled using the following quadratic equation: $V(t) = I - (t - h)^2$, where I and h are unknown constants. The initial height of the ball at time $t = 0$ is 9 ft. and the ball reaches its maximum height of 25 ft. at time $t = 4$. What is the height of the ball at time $t = 2$?

GO ON TO NEXT PAGE

Multiple-Choice

1 (A)(B)(C)(D)(E)
2 (A)(B)(C)(D)(E)
3 (A)(B)(C)(D)(E)
4 (A)(B)(C)(D)(E)
5 (A)(B)(C)(D)(E)
(A)(B)(C)(D)(E)
(A)(B)(C)(D)(E)
(A)(B)(C)(D)(E)
(A)(B)(C)(D)(E)
(A)(B)(C)(D)(E)

Student-Produced Responses

CHAPTER
1 - 7
CUMULATIVE REVIEW

---

1. Amy's age is 4 times Becky's age. In 3 years, Amy will be 3 times as old as Becky. How old is Amy?

(A) 48
(B) 24
(C) 20
(D) 18
(E) 15

---

2. 40/x equals 10 percent of x. Which could be x?

(A) 48
(B) 24
(C) 20
(D) 18
(E) 15

---

3. A = y + 1 and B = x - 2. If x is 5 more than y, A is how much less than B?

(A) 2
(B) 4
(C) 6
(D) 8
(E) 10

---

4. (-1, 5) and (7, 1) are two opposite vertices of a square. What is the area of the square?

(A) 64
(B) 42
(C) 40
(D) 36
(E) 24

---

5. If the lines y = 3x + 4 and 9y = 2x + 1 intersect at the point (a, b), which of the following is true?

(A) a + b = 1
(B) 2a - b = 1
(C) 2a + b = 1
(D) a - b = 1
(E) 2b - a = 1

---

6. The difference of the square of a number and the cube of the same number is 48. What is the square root of the number?

---

7. Given that $4^x = 10$, what is the value of the expression 2^{2x-2}?

---

8. If a mathematics professor can write a complicated math problem in 2 minutes and his teacher's assistant can write a complicated math problem in 3 minutes, how long does it take the two of them working together to write a 10-question quiz?

STOP

Chapter 8: 2D Geometry

6 Sections
52 Practice Questions

8.1 Angles

Important Notes About Angles

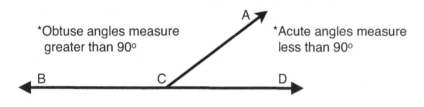

*Obtuse angles measure greater than 90°

*Acute angles measure less than 90°

* Supplementary angles are angles that sum to 180°. Angles BCA and ACD are supplementary.

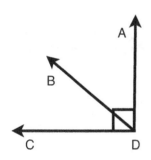

* A right angle is an angle that measures exactly 90°. It is denoted by a square in the corner of the angle.

* Complementary angles are two acute angles that sum to a total of 90°. Angles CDB and BDA are complementary angles.

* Transversals are lines that cut through two parallel lines. Remember your transversal properties:

The following are congruent angles:
Alt. Interior: c & f , d & e
Alt. Exterior: a & h , b & g
Corresponding: a & e , d & h , etc.
Vertical: b & c , e & h , etc.

The following are supplementary angles:
Same Side Int.: c & e , d & f
Same Side Ext.: a & g , b & h

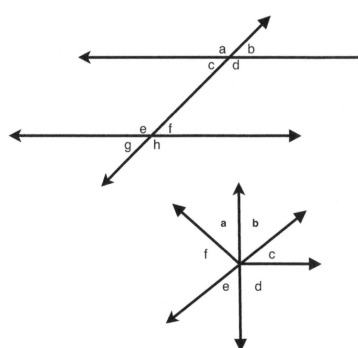

* All of the angles around a single point sum to 360°. So, $a + b + c + d + e + f = 360°$.

* **Important Polygon Equation:**
The sum of all of the interior angles of a polygon with n sides is: **180(n - 2)**

8 ☆ ☆ ☆   Unauthorized copying or
reuse of any part of this
page is illegal.   ☆ ☆ ☆ 8

SAT Example and Technique Application:

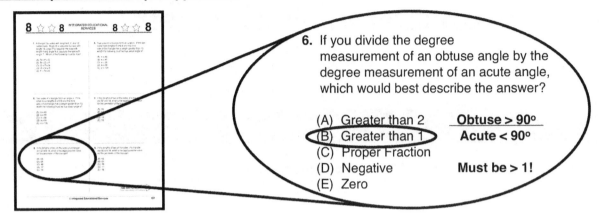

6. If you divide the degree measurement of an obtuse angle by the degree measurement of an acute angle, which would best describe the answer?

(A) Greater than 2   **Obtuse > 90°**
(B) Greater than 1   **Acute < 90°**
(C) Proper Fraction
(D) Negative   **Must be > 1!**
(E) Zero

8.1 **Let's Practice:**

1. Ⓐ Ⓑ Ⓒ Ⓓ Ⓔ   6. Ⓐ Ⓑ Ⓒ Ⓓ Ⓔ
2. Ⓐ Ⓑ Ⓒ Ⓓ Ⓔ   7. Ⓐ Ⓑ Ⓒ Ⓓ Ⓔ
3. Ⓐ Ⓑ Ⓒ Ⓓ Ⓔ   8. Ⓐ Ⓑ Ⓒ Ⓓ Ⓔ
4. Ⓐ Ⓑ Ⓒ Ⓓ Ⓔ   9. Ⓐ Ⓑ Ⓒ Ⓓ Ⓔ
5. Ⓐ Ⓑ Ⓒ Ⓓ Ⓔ   10. Ⓐ Ⓑ Ⓒ Ⓓ Ⓔ

1. In the diagram below, what is the value of $a \div b$?

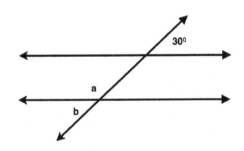

30°

a

b

(A) 1/5
(B) 1/3
(C) 1
(D) 3
(E) 5

2. In the diagram below, what is the value of angle A?

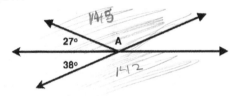

145

27° A

38°

142

(A) 25°
(B) 35°
(C) 115°
(D) 125°
(E) 135°

---

3. In the diagram below, ray OB bisects angle AOC and ray OD bisects angle BOE. If angle COD is 15° and angle AOE is 120°, what is the measurement of angle BOC?

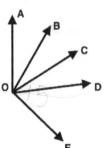

(A) 15
(B) 30
(C) 45
(D) 60
(E) 90

GO ON TO NEXT PAGE ➢

8 ☆ ☆ ☆   Unauthorized copying or
                 reuse of any part of this
                 page is illegal.   ☆ ☆ ☆ 8

8.2 Triangles

8.2.1 Rules of Triangles

Rule 1: **The Third Side Rule:** If given two sides of a triangle, *a* and *b*, the third side, *c*, must meet the following criteria: la - bl < c < a + b

Rule 2: The longest side of a triangle is opposite the largest angle. Likewise, the smallest side is opposite the smallest angle.

SAT Example and Technique Application:

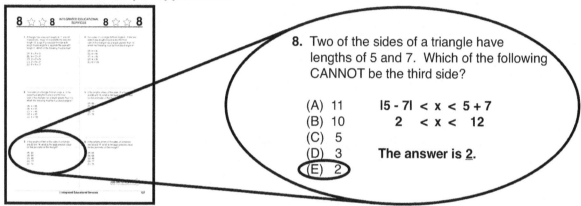

8. Two of the sides of a triangle have lengths of 5 and 7. Which of the following CANNOT be the third side?

(A) 11   l5 - 7l < x < 5 + 7
(B) 10   2   < x <   12
(C) 5
(D) 3   **The answer is 2.**
(E) 2

8.2.1 Let's Practice:

1. Ⓐ Ⓑ Ⓒ Ⓓ Ⓔ   6. Ⓐ Ⓑ Ⓒ Ⓓ Ⓔ
2. Ⓐ Ⓑ Ⓒ Ⓓ Ⓔ   7. Ⓐ Ⓑ Ⓒ Ⓓ Ⓔ
3. Ⓐ Ⓑ Ⓒ Ⓓ Ⓔ   8. Ⓐ Ⓑ Ⓒ Ⓓ Ⓔ
4. Ⓐ Ⓑ Ⓒ Ⓓ Ⓔ   9. Ⓐ Ⓑ Ⓒ Ⓓ Ⓔ
5. Ⓐ Ⓑ Ⓒ Ⓓ Ⓔ   10. Ⓐ Ⓑ Ⓒ Ⓓ Ⓔ

1. A triangle has sides with lengths 6, 7, and 12, respectively. Angle O is opposite the side with length 12, angle P is opposite the side with length 6 and angle N is opposite the side with length 7. Which of the following must be true?

(A) N < P < O
(B) N < O < P
(C) O < P < N
(D) O < N < P
(E) P < N < O

2. Two sides of a triangle form an angle x. If the two sides have lengths 6 and 8 and the third side of this triangle has a length greater than 10, which the following must be true about angle x?

(A) x < 45
(B) x < 90
(C) x = 90
(D) x > 90
(E) x > 135

3. If the lengths of two of the sides of a triangle are 32 and 16, what is the least possible integer value for the perimeter of the triangle?

(A) 65
(B) 64
(C) 48
(D) 17
(E) 16

GO ON TO NEXT PAGE

8 ☆ ☆ ☆

Unauthorized copying or
reuse of any part of this
page is illegal.

☆ ☆ ☆ 8

8.2.2 The Right Triangle

Pythagorean Theorem

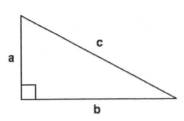

In a right triangle, the side lengths are related to each other through the Pythagorean Theorem. This theorem states that the sum of the squares of the two legs is equivalent to the square of the hypotenuse:

$$a^2 + b^2 = c^2$$

Special Right Triangles (Based on Side Length)

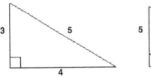 

Certain sets of right triangle side lengths work out so that all sides are of integer lengths. The two triangles above and triangles with proportional side lengths are special triangles that the SAT utilize often in its questions.

Special Right Triangles (Based on Angles)

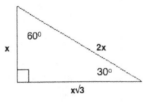

The two most important special right triangles are shown above. The 30-60-90 and 45-45-90 triangles will be used multiple times on any given SAT.

SAT Example and Technique Application:

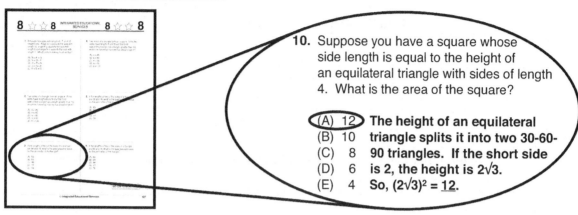

10. Suppose you have a square whose side length is equal to the height of an equilateral triangle with sides of length 4. What is the area of the square?

(A) 12
(B) 10
(C) 8
(D) 6
(E) 4

The height of an equilateral triangle splits it into two 30-60-90 triangles. If the short side is 2, the height is 2√3. So, (2√3)² = 12.

8.2.2 Let's Practice:

1. Ⓐ Ⓑ Ⓒ Ⓓ Ⓔ   6. Ⓐ Ⓑ Ⓒ Ⓓ Ⓔ
2. Ⓐ Ⓑ Ⓒ Ⓓ Ⓔ   7. Ⓐ Ⓑ Ⓒ Ⓓ Ⓔ
3. Ⓐ Ⓑ Ⓒ Ⓓ Ⓔ   8. Ⓐ Ⓑ Ⓒ Ⓓ Ⓔ
4. Ⓐ Ⓑ Ⓒ Ⓓ Ⓔ   9. Ⓐ Ⓑ Ⓒ Ⓓ Ⓔ
5. Ⓐ Ⓑ Ⓒ Ⓓ Ⓔ   10. Ⓐ Ⓑ Ⓒ Ⓓ Ⓔ

1. What is a + b?

(A) 15
(B) 10√3
(C) 10
(D) 5√3
(E) 5

2. What is the area of the triangle below?

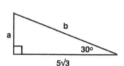

(A) (1/6)√3
(B) (9/2)√3
(C) 6√3
(D) 9√3
(E) 18

3. What is the value of b?

(A) 3
(B) 3√2
(C) 3√3
(D) 6√2
(E) 6√3

4. What is the value of c + d?

(A) 4
(B) 4√2
(C) 8
(D) 8√2
(E) 8 + 4√2

5. What is the value of e + f?

(A) 2
(B) 2√2
(C) 4
(D) 4√2
(E) 8 + 4√2

GO ON TO NEXT PAGE

8.2.3 Additional Triangle Information

Area Equation for Equilateral Triangles

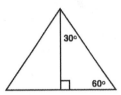

An equilateral triangle can be broken into two 30-60-90 triangles. This allows
you to find the height, so that you can then find the area of the triangle.
However, there is an area formula for equilateral triangles that you can use:

$$\text{Area} = \frac{s^2\sqrt{3}}{4}$$

<u>Additional Note:</u> Any square can be broken into 4 *isosceles* right triangles by
drawing a diagonal through it. (45-45-90)

<u>SAT Example and Technique Application:</u>

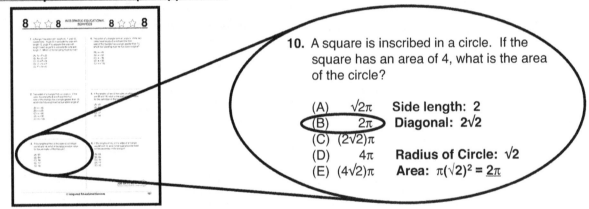

10. A square is inscribed in a circle. If the
square has an area of 4, what is the area
of the circle?

(A) $\sqrt{2}\pi$ **Side length: 2**
(B) 2π **Diagonal: $2\sqrt{2}$**
(C) $(2\sqrt{2})\pi$
(D) 4π **Radius of Circle: $\sqrt{2}$**
(E) $(4\sqrt{2})\pi$ **Area: $\pi(\sqrt{2})^2 = 2\pi$**

8.2.3 <u>Let's Practice:</u>

1. Ⓐ Ⓑ Ⓒ Ⓓ Ⓔ   6. Ⓐ Ⓑ Ⓒ Ⓓ Ⓔ
2. Ⓐ Ⓑ Ⓒ Ⓓ Ⓔ   7. Ⓐ Ⓑ Ⓒ Ⓓ Ⓔ
3. Ⓐ Ⓑ Ⓒ Ⓓ Ⓔ   8. Ⓐ Ⓑ Ⓒ Ⓓ Ⓔ
4. Ⓐ Ⓑ Ⓒ Ⓓ Ⓔ   9. Ⓐ Ⓑ Ⓒ Ⓓ Ⓔ
5. Ⓐ Ⓑ Ⓒ Ⓓ Ⓔ   10. Ⓐ Ⓑ Ⓒ Ⓓ Ⓔ

1. If a rectangle is 10 inches wide and is as tall as
an equilateral triangle with side lengths of 6
inches, what is the area of the rectangle?

(A) 10
(B) $10\sqrt{3}$
(C) 30
(D) $30\sqrt{3}$
(E) 50

2. The height of an equilateral triangle is
equivalent to the diameter of a circle. If the
area of the circle is 12π, what is the height
of the triangle?

(A) 3
(B) $2\sqrt{3}$
(C) $3\sqrt{3}$
(D) $4\sqrt{2}$
(E) $4\sqrt{3}$

GO ON TO NEXT PAGE ⇨

8.2.4 Similar Triangles

Similar triangles are triangles that have all angle measures equivalent and have all sides in proportion to one another. So, if two triangles have the same angles yet are not congruent, the triangles are at least similar.

Demonstration Example

Demo: In the triangle above, a parallel line segment cuts through the larger triangle, creating a smaller triangle. This makes the two triangles similar. What is the value of x + y?

Prop. (x): $\dfrac{6}{9} = \dfrac{6+2}{x}$ --> $6x = 72$ --> $\underline{x = 12}$

Prop. (y): $\dfrac{6}{2} = \dfrac{4}{y}$ --> $6y = 8$ --> $\underline{y = 4/3}$

$x + y = 12 + 4/3 = \underline{40/3}$

SAT Example and Technique Application:

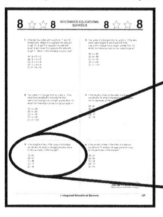

8. A triangle has side lengths 6, 8, and 10. A similar triangle has side lengths 9, 12, and x. What is the value of x?

(A) 15   Sides are proportional:
(B) 16
(C) 17 $\dfrac{6}{10} = \dfrac{9}{x}$ --> $6x = 90$ --> $\underline{x = 15}$
(D) 18
(E) 19

8.2.4 Let's Practice:

1. Ⓐ Ⓑ Ⓒ Ⓓ Ⓔ   6. Ⓐ Ⓑ Ⓒ Ⓓ Ⓔ
2. Ⓐ Ⓑ Ⓒ Ⓓ Ⓔ   7. Ⓐ Ⓑ Ⓒ Ⓓ Ⓔ
3. Ⓐ Ⓑ Ⓒ Ⓓ Ⓔ   8. Ⓐ Ⓑ Ⓒ Ⓓ Ⓔ
4. Ⓐ Ⓑ Ⓒ Ⓓ Ⓔ   9. Ⓐ Ⓑ Ⓒ Ⓓ Ⓔ
5. Ⓐ Ⓑ Ⓒ Ⓓ Ⓔ   10. Ⓐ Ⓑ Ⓒ Ⓓ Ⓔ

1. If a 30-60-90 triangle has a hypotenuse that measures 6, what is the measure of the hypotenuse of a similar triangle that has a shorter leg that measures 2?

(A) 2
(B) 2√3
(C) 4
(D) 4√3
(E) 6

GO ON TO NEXT PAGE ▷

2. What is the value of x?

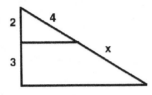

(A) 4
(B) 6
(C) 8
(D) 9
(E) 10

3. What is the value of y/z?

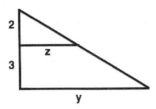

(A) 5/2
(B) 5/3
(C) 3/2
(D) 3/4
(E) 2/3

8.2 Mixed Practice:

1. Ⓐ Ⓑ Ⓒ Ⓓ Ⓔ   6. Ⓐ Ⓑ Ⓒ Ⓓ Ⓔ
2. Ⓐ Ⓑ Ⓒ Ⓓ Ⓔ   7. Ⓐ Ⓑ Ⓒ Ⓓ Ⓔ
3. Ⓐ Ⓑ Ⓒ Ⓓ Ⓔ   8. Ⓐ Ⓑ Ⓒ Ⓓ Ⓔ
4. Ⓐ Ⓑ Ⓒ Ⓓ Ⓔ   9. Ⓐ Ⓑ Ⓒ Ⓓ Ⓔ
5. Ⓐ Ⓑ Ⓒ Ⓓ Ⓔ   10. Ⓐ Ⓑ Ⓒ Ⓓ Ⓔ

1. What is the value of x + y?

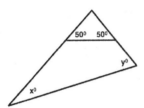

(A) 50
(B) 80
(C) 100
(D) 120
(E) 130

2. What is the area of the triangle below?

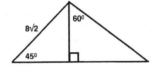

(A) 4√3
(B) 8√3
(C) 8√3 + 8
(D) 32√3 + 32
(E) 64√3 + 64

3. What is the area of the triangle below?

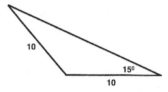

(A) 25
(B) 25√3
(C) 50
(D) 50√3
(E) 100

4. Suppose that AC measures 16, AE measures 8, and BD measures 9. What is the length of BC?

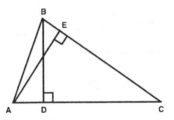

(A) 8√2
(B) 16
(C) 16√3
(D) 18
(E) 18√3

GO ON TO NEXT PAGE

8.3 Circles

There are two important formulas that you must commit to memory when working with circles:

$$A = \pi r^2 \qquad C = 2\pi r = \pi d$$

For most multiple-choice questions, the answers are given in terms of π. There is usually no need to estimate your answers.

Many problems will ask you to work with part of a circle, whether it is an <u>arc</u> or a <u>sector</u>. It is important to remember that an <u>arc</u> is just "a part of the circumference" and a <u>sector</u> is just "a part of the area." This is a further reason why the equations above are so important to memorize. In order to get the part of the circle, just take the given angle, θ, and divide it by 360^0:

Arc Length $= \theta / 360 \ (2\pi r)$ **Area of a Sector $= \theta / 360 \ (\pi r^2)$**

Demonstration Example

<u>Demo</u>: A circle with the center O intersects the square ABCO at the points A and C. If the area of the square is 100, what is the area of the portion of the square that lies outside of the circle?

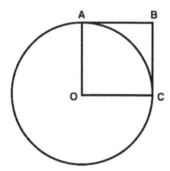

**In order to answer the question, you need a good plan.
First, figure out how you will find the area of the portion of the square outside the circle in terms of the shapes themselves:**

Square - Sector = Area

Then, using the square's area (which you know), the square's side length of 10 (which you can easily figure out), and the sector formula:

$100 - (90/360)(\pi(10)^2) = \underline{100 - 25\pi}$

SAT Example and Technique Application:

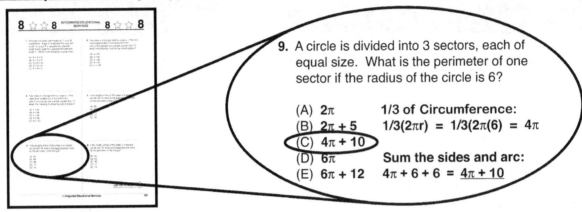

9. A circle is divided into 3 sectors, each of equal size. What is the perimeter of one sector if the radius of the circle is 6?

(A) 2π
(B) $2\pi + 5$
(C) $4\pi + 10$
(D) 6π
(E) $6\pi + 12$

1/3 of Circumference:
$1/3(2\pi r) = 1/3(2\pi(6)) = 4\pi$

Sum the sides and arc:
$4\pi + 6 + 6 = \underline{4\pi + 10}$

8.3 **Let's Practice:**

1. Ⓐ Ⓑ Ⓒ Ⓓ Ⓔ   6. Ⓐ Ⓑ Ⓒ Ⓓ Ⓔ
2. Ⓐ Ⓑ Ⓒ Ⓓ Ⓔ   7. Ⓐ Ⓑ Ⓒ Ⓓ Ⓔ
3. Ⓐ Ⓑ Ⓒ Ⓓ Ⓔ   8. Ⓐ Ⓑ Ⓒ Ⓓ Ⓔ
4. Ⓐ Ⓑ Ⓒ Ⓓ Ⓔ   9. Ⓐ Ⓑ Ⓒ Ⓓ Ⓔ
5. Ⓐ Ⓑ Ⓒ Ⓓ Ⓔ   10. Ⓐ Ⓑ Ⓒ Ⓓ Ⓔ

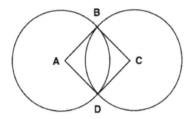

Use the above diagram of circles A and C to answer questions 1 and 2.

1. If ABCD is a square with area 16, what is the perimeter of the central football-shaped region?

(A) 2π
(B) 4π
(C) 6π
(D) 8π
(E) 10π

2. If ABCD is a square with area 16, what is the area of the central football-shaped region?

(A) $\pi - 4$
(B) $4\pi - 8$
(C) $8\pi - 16$
(D) 8π
(E) 16π

Use the diagram above to answer questions 3 and 4.

3. If the emblem shown above is comprised of three circles with radii are in a ratio of 1:2:3 and the largest circle has an area of 81π, what is the area of the black region?

(A) 9π
(B) 16π
(C) 25π
(D) 36π
(E) 45π

4. If the white circles are actually holes cut into the the large black circle and the area of the large black circle is 81π, what is the perimeter of the entire figure, including interior and exterior circles?

(A) 9π
(B) 16π
(C) 25π
(D) 36π
(E) 45π

GO ON TO NEXT PAGE ⟶

8.4 Multiple Shapes in 2 Dimensions

Whenever you are asked to evaluate a form that is not a circle, a rectangle, or a triangle, such a form can usually be broken into pieces that are familiar figures. We will refer to this as "Shape Mathematics." Look at your target shape or distance and see if you can arrive at it by using familiar shapes. Then convert those shapes to familiar equations and proceed.

Demonstration Example

Demo: A circle is inscribed within a square with a side length of 10. What is the area of the portion of the square that is outside of the circle?

Think shapes first:

Then, convert the shapes to equations and solve:

$$\text{Area} = s^2 - \pi r^2 = (10)^2 - \pi(5)^2 = \underline{100 - 25\pi}$$

8.4 Let's Practice:

1. Ⓐ Ⓑ Ⓒ Ⓓ Ⓔ   6. Ⓐ Ⓑ Ⓒ Ⓓ Ⓔ
2. Ⓐ Ⓑ Ⓒ Ⓓ Ⓔ   7. Ⓐ Ⓑ Ⓒ Ⓓ Ⓔ
3. Ⓐ Ⓑ Ⓒ Ⓓ Ⓔ   8. Ⓐ Ⓑ Ⓒ Ⓓ Ⓔ
4. Ⓐ Ⓑ Ⓒ Ⓓ Ⓔ   9. Ⓐ Ⓑ Ⓒ Ⓓ Ⓔ
5. Ⓐ Ⓑ Ⓒ Ⓓ Ⓔ   10. Ⓐ Ⓑ Ⓒ Ⓓ Ⓔ

1. A circle with area 25π is inscribed in a square. What is the perimeter of the square?

   (A) 10
   (B) 25
   (C) 40
   (D) 50
   (E) 75

2. A square with an area of 72 is inscribed inside a circle. What is the circumference of the circle?

   (A) 8π
   (B) 12π
   (C) 16π
   (D) 18π
   (E) 24π

3. A triangle with an area of 18 is shown in the diagram below. If the triangle is isosceles, right, and has a vertex at the center of the circle, what is the area of the portion of the triangle outside the circle?

   (A) $18 - 9/2\pi$
   (B) $18 - 9\pi$
   (C) $36 - 9/2\pi$
   (D) $36 - 9\pi$
   (E) $45 - 6\pi$

4. A semicircle sits atop a square with an area of 64. If the semicircle has the same width as the square, what is the perimeter of the combined figure?

   (A) $8 + 2\pi$
   (B) 8π
   (C) $8 + 4\pi$
   (D) $16 + 2\pi$
   (E) $24 + 4\pi$

GO ON TO NEXT PAGE ▷

Unauthorized copying or
reuse of any part of this
page is illegal.

☆ ☆ ☆ 8

8.5 Geometric Probability

Probability is a concept that we will cover in depth in Chapter 10. However, geometric probability is a topic that we should touch upon now. Geometric probability should be approached the same way that all probability problems are approached. The general rule about probability is that we are looking for <u>favorable outcomes</u>, what we "want," over <u>total outcomes</u>, what is "possible." Try to apply this concept to the questions below:

Practice Problem 1:

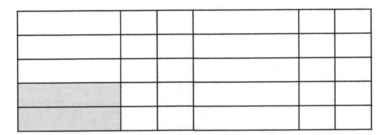

The grid above is made up of repeating rectangles. Each of the larger rectangles has a horizontal length that is three times the length of the smaller rectangles and all of the rectangles are the same height. If a dart is thrown at random and definitely hits the grid, what is the probability that the dart will land in the shaded region?

Practice Problem 2:

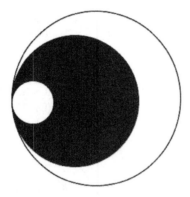

The three circles above share one common point at the left. The circles' diameters are related in a ratio of 1:3:4. If a point within the largest circle is picked at random, what is the probability that this point will be in the shaded region?

8.6 Covering an Area

You will often be asked to figure out how many identical objects or units of a certain area can be fit inside or placed over a larger area. Often times, you can figure this out by dividing the large area by the area of one of the small repeating units.

SAT Example and Technique Application:

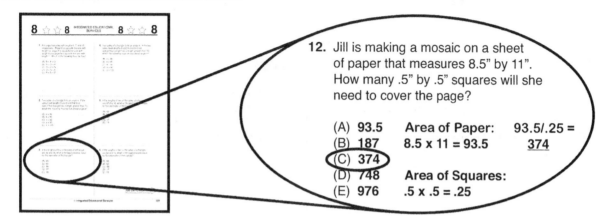

12. Jill is making a mosaic on a sheet of paper that measures 8.5" by 11". How many .5" by .5" squares will she need to cover the page?

(A) 93.5   **Area of Paper:**   93.5/.25 =
(B) 187   8.5 x 11 = 93.5   __374__
(C) 374
(D) 748   **Area of Squares:**
(E) 976   .5 x .5 = .25

8.6 Let's Practice:

1. Ⓐ Ⓑ Ⓒ Ⓓ Ⓔ   6. Ⓐ Ⓑ Ⓒ Ⓓ Ⓔ
2. Ⓐ Ⓑ Ⓒ Ⓓ Ⓔ   7. Ⓐ Ⓑ Ⓒ Ⓓ Ⓔ
3. Ⓐ Ⓑ Ⓒ Ⓓ Ⓔ   8. Ⓐ Ⓑ Ⓒ Ⓓ Ⓔ
4. Ⓐ Ⓑ Ⓒ Ⓓ Ⓔ   9. Ⓐ Ⓑ Ⓒ Ⓓ Ⓔ
5. Ⓐ Ⓑ Ⓒ Ⓓ Ⓔ   10. Ⓐ Ⓑ Ⓒ Ⓓ Ⓔ

1. Beth wants to tile her bathroom floor. She chooses a tile pattern made of 4 differently-colored square tiles in which the 4 tiles are put together to make a larger square. This pattern is used to cover a small bathroom floor that is 4 feet by 6 feet. If each small tile is 2 inches by 2 inches, how many repetitions of the tile pattern will it take to cover the entire bathroom floor?

(A) 216
(B) 432
(C) 864
(D) 1728
(E) 3456

2. A kindergarten teacher is making name tags for her class. Each name tag is 4 by 6 inches and there are 18 students in the class. If the teacher is cutting these tags from 8 by 10 inch pieces of construction paper, what is the least number of sheets of construction paper she will need to make tags for the whole class?

(A) 6
(B) 8
(C) 9
(D) 10
(E) 18

3. A square with its diagonals is shown below. If the area of the shaded triangle is 7, what is the area of the square?

(A) 28
(B) 42
(C) 56
(D) 70
(E) 84

GO ON TO NEXT PAGE ⟩

8 ☆ ☆ ☆

Unauthorized copying or reuse of any part of this page is illegal.

☆ ☆ ☆ 8

CHAPTER 8: CHALLENGE QUESTIONS

Student-Produced Responses

8.1

1. 15 angles have consecutive measurements and, when placed adjacently, sum to 180 degrees. What is the measure of the largest angle minus the measure of the smallest angle?

8.2

2. The hypotenuse of a 30-60-90 right triangle measures $x^2 - 8$ and the shortest leg measures x. What is the length of the hypotenuse?

3. An equilateral triangle has sides of length $x - 4$. If the area of the triangle is $\sqrt{3}/2(x)$, what is a possible value for the side length of the triangle?

4. Two triangles share a common angle and have parallel bases. The common angle is not adjacent to the parallel bases. If the height of one of the triangles is 8 and the length of its base is 9, what is the height of the other triangle if the length of its base is 15?

8.3

5. Four congruent circles are arranged 2 by 2 and inscribed in a square that has an area of 128. If these circles are then inscribed in a larger circle, what is the radius of the larger circle? *(Answer in decimal form.)*

8.4

6. Suppose that an equilateral triangle is inscribed within a circle. If the side length of the triangle is 5, what is the area of the portion of the circle that lies outside the triangle? *(Answer in decimal form.)*

8.5

7. If the diameters of the small, medium, and large circles in the figure below are related in a ratio of 1:2:3, what is the probability that a point in the unshaded region will be randomly selected?

GO ON TO NEXT PAGE →

8 ☆ ☆ ☆   ☆ ☆ ☆ 8

Unauthorized copying or
reuse of any part of this
page is illegal.

Multiple-Choice   **Student-Produced Responses**

CHAPTER
8
REVIEW

1 Ⓐ Ⓑ Ⓒ Ⓓ Ⓔ
2 Ⓐ Ⓑ Ⓒ Ⓓ Ⓔ
3 Ⓐ Ⓑ Ⓒ Ⓓ Ⓔ
4 Ⓐ Ⓑ Ⓒ Ⓓ Ⓔ
Ⓐ Ⓑ Ⓒ Ⓓ Ⓔ
Ⓐ Ⓑ Ⓒ Ⓓ Ⓔ
Ⓐ Ⓑ Ⓒ Ⓓ Ⓔ
Ⓐ Ⓑ Ⓒ Ⓓ Ⓔ
Ⓐ Ⓑ Ⓒ Ⓓ Ⓔ
Ⓐ Ⓑ Ⓒ Ⓓ Ⓔ

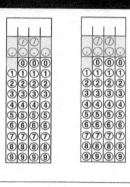  

5   6

---

1. A ladder is leaning against a house and the base of the ladder is 5 feet from the house. If the ladder touches the house 12 feet up, what is the length of the ladder?

   (A) 10
   (B) 12
   (C) 13
   (D) 15
   (E) 16

---

2. A tree has fallen and is leaning against another tree, forming a 30⁰ angle with the ground. If the fallen tree is 58 ft. tall, how high up does the fallen tree touch the standing tree?

   (A) 24 ft.
   (B) 25 ft.
   (C) 29 ft.
   (D) 30 ft.
   (E) 32 ft.

---

3. A circle is centered at the origin and contains the point (12,9). What is the area of the circle?

   (A) 64π
   (B) 121π
   (C) 169π
   (D) 196π
   (E) 225π

---

4. If a circle is centered at the origin and tangent to the line y = -x + 4, what is the area of the circle divided by π?

   (A) 2
   (B) 4
   (C) 8
   (D) 12
   (E) 16

---

5. If the equilateral triangle pictured below has an area of 9√3, is centered at the origin, and BC is parallel to the x-axis, what is the sum of the absolute values of the x-coordinates of B and C?

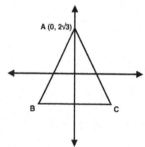

A (0, 2√3)

---

6. In the diagram below, what is the radius of the central circle if the area of the square is 64? *(Answer in decimal form.)*

GO ON TO NEXT PAGE ⟫

8 ☆ ☆ ☆

Unauthorized copying or reuse of any part of this page is illegal.

☆ ☆ ☆ 8

| Multiple-Choice | Student-Produced Responses |

CHAPTER
1 - 8
CUMULATIVE REVIEW

1 Ⓐ Ⓑ Ⓒ Ⓓ Ⓔ
2 Ⓐ Ⓑ Ⓒ Ⓓ Ⓔ
3 Ⓐ Ⓑ Ⓒ Ⓓ Ⓔ
4 Ⓐ Ⓑ Ⓒ Ⓓ Ⓔ
Ⓐ Ⓑ Ⓒ Ⓓ Ⓔ
Ⓐ Ⓑ Ⓒ Ⓓ Ⓔ
Ⓐ Ⓑ Ⓒ Ⓓ Ⓔ
Ⓐ Ⓑ Ⓒ Ⓓ Ⓔ
Ⓐ Ⓑ Ⓒ Ⓓ Ⓔ
Ⓐ Ⓑ Ⓒ Ⓓ Ⓔ

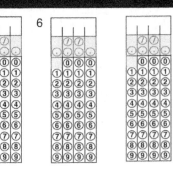
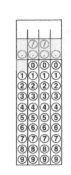

5   6

1. If $-2 < x < -1$, which of the following has the least value?

   (A) $-x$
   (B) $1/x$
   (C) $-1/x$
   (D) $1/x^2$
   (E) $-1/x^2$

2. Given that $|x + 3| < |x - 3|$, the entire solution set for x includes all values of x which are:

   (A) Less than 3
   (B) Greater than 3
   (C) Less than -3
   (D) Greater than -3
   (E) Less than 0

3. What is the greatest possible number of Januarys that could occur in a ten-year period?

   (A) 9
   (B) 10
   (C) 11
   (D) 12
   (E) 13

4. If $2x - 7 = 12$, what is the value of $4x + 1$?

   (A) 41
   (B) 39
   (C) 38
   (D) 29
   (E) 28

5. Let the operations ◇ and ★ be defined for all real numbers x and y as follows:

$$x \diamond y = x + 2y$$
$$x \star y = 2x - 5y$$

   If $6 \diamond (3y) = (8y) \star 2$, what is the value of y?

6. The parabola pictured below is the graph of $y = (1/3)x^2 - 3x + 6$. What is the area of the rectangle in the diagram?

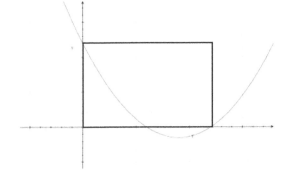

STOP

Chapter 9: 3D Geometry
9 Sections
56 Practice Questions

9.1 Overview of Solid Geometry

Any **uniform solid or prism** can be thought of as a flat stack of objects, or as a group of bases piled to a certain height. For example, a cylinder is a stack of circles that is h units high. This gives us an easy way of thinking about volume. For any uniform or regular prism, the volume is Base x Height. For any solid with a single pointed tip, it is V = 1/3(Base)(Height).

B = Base Area	V = Base x Height	V = 1/3(Base)(Height)
Rectangle = LW	Rectangular Prism = LWH	
Circular Base = πr^2	Cylinder = $\pi r^2 h$	Cone = 1/3 $\pi r^2 h$
Square = s^2	Cube = s^3	4 Sided Pyramid = 1/3 $s^2 h$

As long as you think about volume in this way, you can remember the above formulas very easily by keeping this table in mind. You can also find **surface area** in some cases by simply adding up the areas of all the faces of the solid. Thinking about solid geometry in this way leaves you with fewer formulas to memorize.

Of course, you will have to remember a few facts and formulas, but more importantly you will need to become comfortable applying them. The problems in this chapter are typical examples of the solid geometry questions you will actually see on the SAT, so don't be surprised if you see some of them again.

Here are some additional formulas you will need to memorize, but without using the above method:

Formulas for Spheres: Volume = 4/3 πr^3 and Surface Area = 4 πr^2

Surface Area of a Cylinder: $2\pi r^2 + 2\pi rh$

Surface Area of a Cone: $\pi r^2 + \pi r\sqrt{(r^2 + h^2)}$ (It is unlikely that you will need this.)

3-Dimensional Distance Formula: $d^2 = l^2 + w^2 + h^2$

9.2 Cubes

Demonstration Example

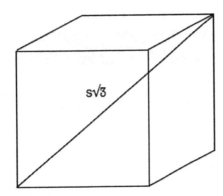

$s\sqrt{3}$

Formulas:

**Side = ** s

**Surface Area = ** $6s^2$

**Volume = ** s^3

**Longest Diagonal = ** $s\sqrt{3}$

SAT Example and Technique Application:

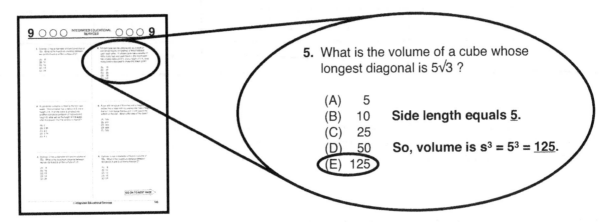

5. What is the volume of a cube whose longest diagonal is $5\sqrt{3}$?

(A)   5
(B)   10   **Side length equals 5.**
(C)   25
(D)   50   **So, volume is $s^3 = 5^3 = 125$.**
(E)   125

9.2 Let's Practice:

1. Ⓐ Ⓑ Ⓒ Ⓓ Ⓔ   6. Ⓐ Ⓑ Ⓒ Ⓓ Ⓔ
2. Ⓐ Ⓑ Ⓒ Ⓓ Ⓔ   7. Ⓐ Ⓑ Ⓒ Ⓓ Ⓔ
3. Ⓐ Ⓑ Ⓒ Ⓓ Ⓔ   8. Ⓐ Ⓑ Ⓒ Ⓓ Ⓔ
4. Ⓐ Ⓑ Ⓒ Ⓓ Ⓔ   9. Ⓐ Ⓑ Ⓒ Ⓓ Ⓔ
5. Ⓐ Ⓑ Ⓒ Ⓓ Ⓔ   10. Ⓐ Ⓑ Ⓒ Ⓓ Ⓔ

1. What is the surface area of a cube that has a volume of 729?

(A)   9
(B)   81
(C)   243
(D)   486
(E)   729

2. A cube-shaped room is carpeted and its walls, but not its ceiling, are painted. If the volume of the room is 125, what is the total surface area of both the painted and carpeted surfaces in the room?

(A)   25
(B)   100
(C)   125
(D)   130
(E)   150

GO ON TO NEXT PAGE ➡

9.3 Rectangular Solids

Demonstration Example

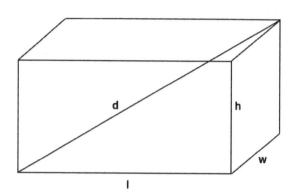

Formulas:

Side = l, w, h

Surface Area = 2(lw + lh + wh)

Volume = lwh

L.D. => $d^2 = l^2 + w^2 + h^2$

SAT Example and Technique Application:

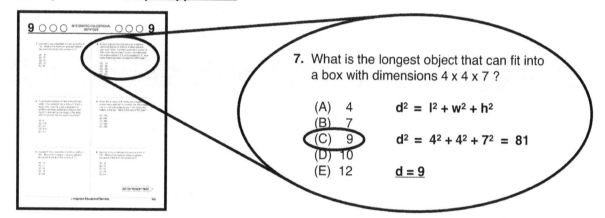

7. What is the longest object that can fit into a box with dimensions 4 x 4 x 7 ?

(A)   4 $d^2 = l^2 + w^2 + h^2$
(B)   7
(C)   9 $d^2 = 4^2 + 4^2 + 7^2 = 81$
(D)   10
(E)   12   <u>d = 9</u>

9.3 Let's Practice:

1. Ⓐ Ⓑ Ⓒ Ⓓ Ⓔ   6. Ⓐ Ⓑ Ⓒ Ⓓ Ⓔ
2. Ⓐ Ⓑ Ⓒ Ⓓ Ⓔ   7. Ⓐ Ⓑ Ⓒ Ⓓ Ⓔ
3. Ⓐ Ⓑ Ⓒ Ⓓ Ⓔ   8. Ⓐ Ⓑ Ⓒ Ⓓ Ⓔ
4. Ⓐ Ⓑ Ⓒ Ⓓ Ⓔ   9. Ⓐ Ⓑ Ⓒ Ⓓ Ⓔ
5. Ⓐ Ⓑ Ⓒ Ⓓ Ⓔ   10. Ⓐ Ⓑ Ⓒ Ⓓ Ⓔ

1. What is the length of the longest stick that will fit inside a box with dimensions 2 x 9 x 6 ?

(A)   9
(B)   11
(C)   12
(D)   15
(E)   16

2. If the area of the top of a rectangular solid is 12, the front has an area of 6, and the side has an area of 2, what is the volume of this solid?

(A)   12
(B)   24
(C)   100
(D)   121
(E)   144

GO ON TO NEXT PAGE ➡

3. A pencil case has dimensions of 6" x 3" x 2". How many of the following pencil sizes can entirely fit inside the case ?

10" , 8" , 7.5" , 7" , 5"

(A) 1
(B) 2
(C) 3
(D) 4
(E) 5

4. A book with dimensions 10" x 12" x 2" has a hard cover. What is the area of the cover?
(The book is oriented to be taller than it is wide.)

(A)   24
(B)   66
(C)   132
(D)   264
(E)   528

5. A rectangular solid has faces A, B, and C. The area of face A is *a*, The area of face B is *b*, and the area of face C is *c*. What is the volume of the rectangular solid in terms of *a*, *b*, and *c*.

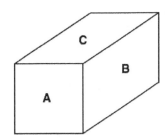

(A) $a\sqrt{bc}$
(B) a^2bc
(C) $a^2b\sqrt{c}$
(D) abc
(E) $\sqrt{abc}$

9.4 Cylinders

Demonstration Example

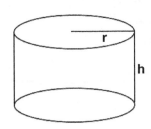

Formulas:

Volume = $\pi r^2 h$

Surface Area = $2\pi r^2 + 2\pi rh$

SAT Example and Technique Application:

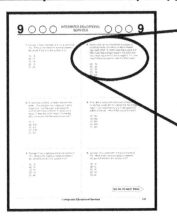

9. A can of tomato soup has a diameter of 4 inches and a height of 10 inches. What is the area of a label that covers the can's entire side?

(A) 10π
(B) 14π
(C) 20π
(D) 40π
(E) 50π

$C = 2\pi r = 2\pi(2) = 4\pi$

$2\pi rh = 4\pi(10) = \underline{40\pi}$

9.4 Let's Practice:

1. Ⓐ Ⓑ Ⓒ Ⓓ Ⓔ   6. Ⓐ Ⓑ Ⓒ Ⓓ Ⓔ
2. Ⓐ Ⓑ Ⓒ Ⓓ Ⓔ   7. Ⓐ Ⓑ Ⓒ Ⓓ Ⓔ
3. Ⓐ Ⓑ Ⓒ Ⓓ Ⓔ   8. Ⓐ Ⓑ Ⓒ Ⓓ Ⓔ
4. Ⓐ Ⓑ Ⓒ Ⓓ Ⓔ   9. Ⓐ Ⓑ Ⓒ Ⓓ Ⓔ
5. Ⓐ Ⓑ Ⓒ Ⓓ Ⓔ   10. Ⓐ Ⓑ Ⓒ Ⓓ Ⓔ

1. Cylinder C has a diameter of 6 and a volume of 72π. What is the maximum distance between two points A and B on the surface of C?

(A)   8
(B) 10
(C) 12
(D) 14
(E) 24

2. A cylindrical container is filled to the brim with water. This container has a radius of 3 and a height of 9. If all the water is emptied into another cylindrical container of radius 6 and height 3, what will be the height of the water after it is poured into the second container?

(A) 2
(B) 2.25
(C) 2.5
(D) 2.75
(E) 4.5

3. A totem pole can be understood as a stack of cylindrical blocks (or totems) of wood stacked upon each other. If a totem pole has a volume of 400π cubic feet and each block in the totem pole has a base radius of 2 ft. and a height of 1 ft., how many totems are used to make this totem pole?

(A)   10
(B)   20
(C)   25
(D)   50
(E) 100

4. A can with a radius of 6 inches and a height of 6 inches has a label with no overlap that has a height that is 1 inch below the top and 1 inch above the bottom of the can. What is the area of the label?

(A) 12π
(B) 24π
(C) 48π
(D) 60π
(E) 72π

9.5 The One Third Rule

The Rule of One Third:

The volume of any shape that tapers to a single point (such as a pyramid or a cone) is always 1/3 of the volume of the corresponding shape that does not taper to a point (such as a prism or a cylinder).

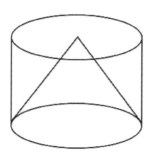

Formulas:

Volume of a Cone = $1/3\pi r^2 h$

Volume of a Square Pyramid = $1/3 s^2 h$

SAT Example and Technique Application:

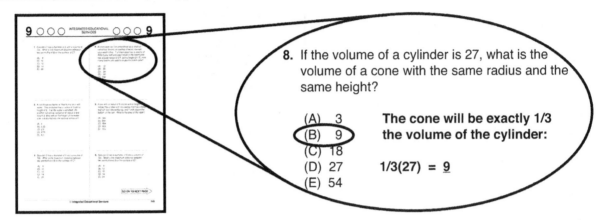

8. If the volume of a cylinder is 27, what is the volume of a cone with the same radius and the same height?

(A) 3
(B) 9
(C) 18
(D) 27
(E) 54

The cone will be exactly 1/3 the volume of the cylinder:

1/3(27) = <u>9</u>

9.5 **Let's Practice:**

1. (A)(B)(C)(D)(E)   6. (A)(B)(C)(D)(E)
2. (A)(B)(C)(D)(E)   7. (A)(B)(C)(D)(E)
3. (A)(B)(C)(D)(E)   8. (A)(B)(C)(D)(E)
4. (A)(B)(C)(D)(E)   9. (A)(B)(C)(D)(E)
5. (A)(B)(C)(D)(E)   10. (A)(B)(C)(D)(E)

1. A sculptor takes a cube of marble with sides of 12 inches and carves a square pyramid out of it. What is the volume of marble that is carved away?

(A) 1728
(B) 1296
(C) 1152
(D) 864
(E) 432

2. A cone with height 9 and diameter 4 is filled with water. If all the water is poured into a cylindrical container with the same dimensions, how deep is the water in the cylindrical container?

(A) 3
(B) 4
(C) 6
(D) 8
(E) 9

3. A piece of cardboard in the shape of an equilateral triangle with sides of length 12 is to be folded along the dotted lines so that the three points meet, creating a pyramid. What is the volume of the resulting pyramid?

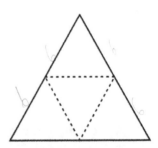

(A) 9√2
(B) 18√2
(C) 18√3
(D) 27
(E) 36

GO ON TO NEXT PAGE ⟹

9.6 Spheres

Demonstration Example

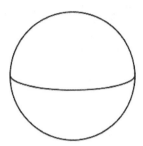

Formulas:

Volume of a Sphere = $4/3\pi r^3$

Surface Area of a Sphere = $4\pi r^2$

SAT Example and Technique Application:

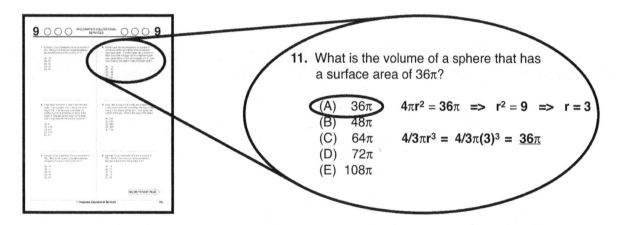

11. What is the volume of a sphere that has a surface area of 36π?

(A) 36π $\quad 4\pi r^2 = 36\pi \implies r^2 = 9 \implies r = 3$
(B) 48π
(C) 64π $\quad 4/3\pi r^3 = 4/3\pi(3)^3 = \underline{36\pi}$
(D) 72π
(E) 108π

9.6 Let's Practice:

1. Ⓐ Ⓑ Ⓒ Ⓓ Ⓔ   6. Ⓐ Ⓑ Ⓒ Ⓓ Ⓔ
2. Ⓐ Ⓑ Ⓒ Ⓓ Ⓔ   7. Ⓐ Ⓑ Ⓒ Ⓓ Ⓔ
3. Ⓐ Ⓑ Ⓒ Ⓓ Ⓔ   8. Ⓐ Ⓑ Ⓒ Ⓓ Ⓔ
4. Ⓐ Ⓑ Ⓒ Ⓓ Ⓔ   9. Ⓐ Ⓑ Ⓒ Ⓓ Ⓔ
5. Ⓐ Ⓑ Ⓒ Ⓓ Ⓔ   10. Ⓐ Ⓑ Ⓒ Ⓓ Ⓔ

1. A sphere encloses a cube with a volume of 1000 so that all vertices of the cube touch the surface of the sphere. What is the volume of the sphere?

(A) $250\pi\sqrt{2}$
(B) $250\pi\sqrt{3}$
(C) $480\pi\sqrt{2}$
(D) $500\pi\sqrt{3}$
(E) 500π

2. A cube-shaped box encloses a sphere. The sphere is tangent to the center of each face of the cube. If the volume of the box is 125, what is the surface area of the sphere?

(A) 50π
(B) 25π
(C) 20π
(D) 10π
(E) 5π

GO ON TO NEXT PAGE ⟩

3. Planet X has a surface area of 400π square kilometers. How long is the equator of Planet X?

(A) 50π
(B) 25π
(C) 20π
(D) 10π
(E) 5π

4. A sphere is kept in a cubic box with sides of length 2 and touches all sides of the box. What is the volume of the sphere?

(A) $4/3\pi$
(B) $2/3\pi$
(C) $1/3\pi$
(D) $1/4\pi$
(E) $1/6\pi$

5. Three spherical tennis balls are kept in a cylindrical can whose sides touch the edges of the balls. If the volume of one of the tennis balls is 36π, what is the volume of the can?

(A) 9π
(B) 72π
(C) 81π
(D) 124π
(E) 162π

9.7 Breaking Up Volumes

Use what you just learned about prisms and solids to answer these questions. In many of them, it will be helpful to think of cross sections (slices taken perpendicular to a solid's base, so that the cross section of a cube is a square and the cross section of a cylinder is a circle, and so on.) You can also figure out how many smaller objects of a set volume can fit into a larger volume by dividing the larger volume by the smaller volume.

Demonstration Example

Demo: Ceramic tiles are piled upon each other to form a cube. Each tile is 4 inches by 4 inches by 1/4 inch thick. How many tiles are there?

In a cube, all the edges are equal and each "slice" or cross section is a square. Each square tile is 4x4, so the height of all the tiles must be 4 as well.

$$\frac{\text{Total Height}}{\text{Individual Height}} = \frac{4 \text{ inches}}{1/4 \text{ in. per tile}} = 16 \text{ tiles}$$

Alternatively, one could find the volume of the stack: 4 x 4 x 4 = 64 cubic inches, and then divide by the volume of each tile: 4 x 4 x 1/4 = 4 cubic inches.

So, 64 / 4 = 16 tiles

SAT Example and Technique Application:

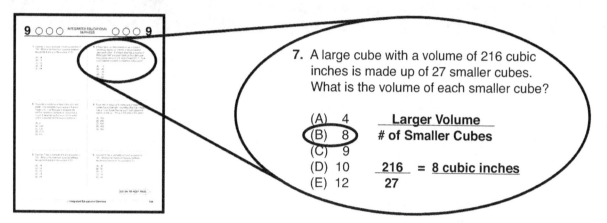

7. A large cube with a volume of 216 cubic inches is made up of 27 smaller cubes. What is the volume of each smaller cube?

(A) 4
(B) 8
(C) 9
(D) 10
(E) 12

$$\frac{\text{Larger Volume}}{\text{\# of Smaller Cubes}}$$

$$\frac{216}{27} = 8 \text{ cubic inches}$$

9.7 **Let's Practice:**

1. Ⓐ Ⓑ Ⓒ Ⓓ Ⓔ   6. Ⓐ Ⓑ Ⓒ Ⓓ Ⓔ
2. Ⓐ Ⓑ Ⓒ Ⓓ Ⓔ   7. Ⓐ Ⓑ Ⓒ Ⓓ Ⓔ
3. Ⓐ Ⓑ Ⓒ Ⓓ Ⓔ   8. Ⓐ Ⓑ Ⓒ Ⓓ Ⓔ
4. Ⓐ Ⓑ Ⓒ Ⓓ Ⓔ   9. Ⓐ Ⓑ Ⓒ Ⓓ Ⓔ
5. Ⓐ Ⓑ Ⓒ Ⓓ Ⓔ   10. Ⓐ Ⓑ Ⓒ Ⓓ Ⓔ

1. A certain uniform prism is 8 units high and each cross section of this prism is an equilateral triangle. If all of the edges of the prism's base are 4 units long, what is the volume of the prism?

(A) 8√3
(B) 16√3
(C) 32√3
(D) 48√3
(E) 64√3

2. The DVDs at a rental store are piled up in a stack. The volume of the stack is 200 inches cubed. Each DVD has a protective sleeve that is 5 inches by 5 inches and is 1/4 inch thick. How many DVDs are in this stack?

(A) 32
(B) 36
(C) 40
(D) 44
(E) 48

3. A block of ice is shaped like a rectangular prism with dimensions of 8 inches x 12 inches x 14 inches. Cubes of ice are cut from the block. If each cube has edges of length 2 inches, how many such cubes can be cut from this block of ice?

(A) 148
(B) 168
(C) 212
(D) 224
(E) 312

4. A roll of candies is arranged as a cylinder with a volume of 18π cubic centimeters. If each candy has a diameter of 3 cm and a height of 1 cm, how many such candies are in each roll?

(A) 4
(B) 5
(C) 6
(D) 7
(E) 8

GO ON TO NEXT PAGE ⟩

9.8 Volume Shift

Volume shift problems involve displacing the volume of one 3-dimensional object into another object. For example, you may be given a rectangular solid that is filled with sand; the SAT will then ask you what the height of the sand will be if all of the sand in this solid is poured into a second rectangular solid. In this case, the given height of the second solid is an irrelevant piece of information. You must assume that the sand will not fill the second solid to the top. So, the height of the sand will be h, or the variable you are attempting to solve for. In this case, you set the volume of the original solid equal to the volume equation for the second solid, in which the height is replaced with a variable. See the example below:

SAT Example and Technique Application:

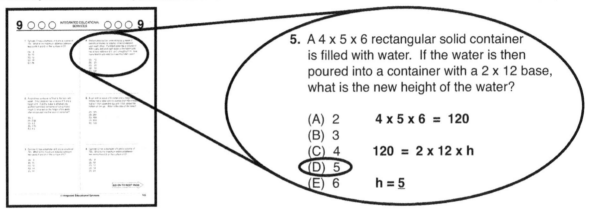

5. A 4 x 5 x 6 rectangular solid container is filled with water. If the water is then poured into a container with a 2 x 12 base, what is the new height of the water?

(A) 2   **4 x 5 x 6 = 120**
(B) 3
(C) 4   **120 = 2 x 12 x h**
(D) 5
(E) 6   **h = 5**

9.8 Let's Practice:

1. Ⓐ Ⓑ Ⓒ Ⓓ Ⓔ   6. Ⓐ Ⓑ Ⓒ Ⓓ Ⓔ
2. Ⓐ Ⓑ Ⓒ Ⓓ Ⓔ   7. Ⓐ Ⓑ Ⓒ Ⓓ Ⓔ
3. Ⓐ Ⓑ Ⓒ Ⓓ Ⓔ   8. Ⓐ Ⓑ Ⓒ Ⓓ Ⓔ
4. Ⓐ Ⓑ Ⓒ Ⓓ Ⓔ   9. Ⓐ Ⓑ Ⓒ Ⓓ Ⓔ
5. Ⓐ Ⓑ Ⓒ Ⓓ Ⓔ   10. Ⓐ Ⓑ Ⓒ Ⓓ Ⓔ

1. A cylindrical flask is filled to the rim with water. This flask has a diameter of 6 and a height of 8. If all of the water is emptied into another cylindrical container of radius 6 and height 8, what will be the height of the water in the second container?

(A) 2
(B) 3
(C) 4
(D) 6
(E) 8

2. A rectangular fish tank has a length of 3 ft., a width of 2.5 ft. and a height of 2 ft. This tank is entirely filled with water before all the water is emptied into a second tank of equal volume. If the second tank has a width of 1 ft. and a length of 5 ft., how tall is the second tank?

(A) 2.5
(B) 3
(C) 4
(D) 4.5
(E) 6.5

GO ON TO NEXT PAGE ⟩

9.9 3D Ratios

9.9.1 **Ratio Conversion Table**

Dimension	Object A	Object B
1D (Length, Width, Circumference)	X	Y
2D (Area, Surface Area)	X^2	Y^2
3D (Volume, Mass)	X^3	Y^3

* **When you are given facts about objects in different dimensions, you can use the table above to convert from one dimension to any other dimension.**

* **Note: When converting from 3D to 2D, it is best to convert down to 1D first, then back up to 2D.**

SAT Example and Technique Application:

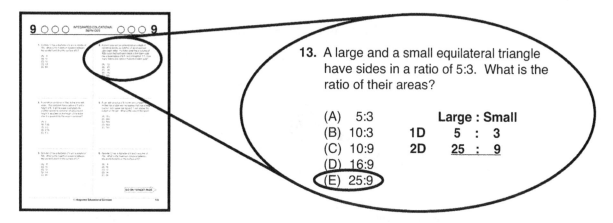

13. A large and a small equilateral triangle have sides in a ratio of 5:3. What is the ratio of their areas?

(A) 5:3
(B) 10:3
(C) 10:9
(D) 16:9
(E) 25:9

	Large : Small
1D	5 : 3
2D	25 : 9

9.9.1 **Let's Practice:**

1. Ⓐ Ⓑ Ⓒ Ⓓ Ⓔ   6. Ⓐ Ⓑ Ⓒ Ⓓ Ⓔ
2. Ⓐ Ⓑ Ⓒ Ⓓ Ⓔ   7. Ⓐ Ⓑ Ⓒ Ⓓ Ⓔ
3. Ⓐ Ⓑ Ⓒ Ⓓ Ⓔ   8. Ⓐ Ⓑ Ⓒ Ⓓ Ⓔ
4. Ⓐ Ⓑ Ⓒ Ⓓ Ⓔ   9. Ⓐ Ⓑ Ⓒ Ⓓ Ⓔ
5. Ⓐ Ⓑ Ⓒ Ⓓ Ⓔ   10. Ⓐ Ⓑ Ⓒ Ⓓ Ⓔ

1. Circle A has an area four times the area of Circle B. What is the ratio of the circumference of Circle A to the circumference of Circle B?

(A) 1:1
(B) 2:1
(C) 4:1
(D) 8:1
(E) 16:1

GO ON TO NEXT PAGE ⟩

2. A large golden sphere is melted down and the liquid gold is used to make 64 identical small spheres. What is the ratio of the surface area of the large sphere to the surface area of one of the small spheres?

(A) 2:1
(B) 4:1
(C) 8:1
(D) 16:1
(E) 64:1

3. The perimeter of square X is three times the perimeter of square Y. What is the ratio of the sides of X and Y?

(A) 1:3
(B) 2:3
(C) 3:2
(D) 3:1
(E) 9:1

4. The length of each side of an equilateral triangle is tripled to create a second triangle. The area of the second triangle is how many times the area of the original triangle?

(A) 3
(B) 6
(C) 9
(D) 12
(E) 27

9.9.2 **Shaded:Unshaded Areas**

Demonstration Example

Demo: What is the ratio of shaded to unshaded areas in the following figure?

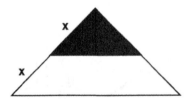

* **If you have two similar figures with common measurements in one dimension, you can use a ratio conversion table to find the ratio of shaded to unshaded regions for figures like the one above.**

	Small	Large
1D	1	2
2D	1	4

Shaded Area = Area of the small triangle = 1
Unshaded Area = Area of the large triangle minus the small = 4 - 1 = 3

Thus, the ratio of the shaded area to the unshaded area is 1:3.

9.9.2 **Let's Practice:**

1. Ⓐ Ⓑ Ⓒ Ⓓ Ⓔ 6. Ⓐ Ⓑ Ⓒ Ⓓ Ⓔ
2. Ⓐ Ⓑ Ⓒ Ⓓ Ⓔ 7. Ⓐ Ⓑ Ⓒ Ⓓ Ⓔ
3. Ⓐ Ⓑ Ⓒ Ⓓ Ⓔ 8. Ⓐ Ⓑ Ⓒ Ⓓ Ⓔ
4. Ⓐ Ⓑ Ⓒ Ⓓ Ⓔ 9. Ⓐ Ⓑ Ⓒ Ⓓ Ⓔ
5. Ⓐ Ⓑ Ⓒ Ⓓ Ⓔ 10. Ⓐ Ⓑ Ⓒ Ⓓ Ⓔ

1. In the picture below, two parallel lines run through the triangle and the left side is cut into three equal parts. What is the ratio of the area of the shaded region to the area of the unshaded regions?

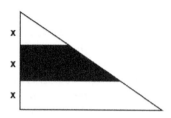

(A) 1:3
(B) 1:2
(C) 1:1
(D) 2:1
(E) 3:1

2. The large circle below has a diameter that is cut into four equal parts. What is the ratio of the shaded area to the unshaded area?

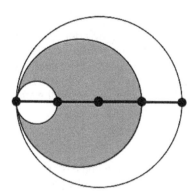

(A) 1:3
(B) 1:2
(C) 1:1
(D) 2:1
(E) 3:1

3. The figure below displays an oblique pyramid with a square base. Two parallel planes cut the sides of the diagonal edges into three equal lengths. What is the ratio of the volume of the middle section (shaded) to the volume of the upper and lower sections (unshaded)?

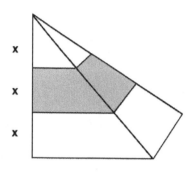

(A) 1:5
(B) 1:3
(C) 7:20
(D) 8:20
(E) 11:30

4. The large circle below has a diameter that is cut into three equal parts. What is the ratio of the shaded area to the unshaded area?

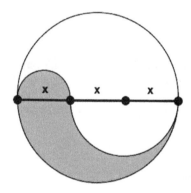

(A) 1:2
(B) 2:3
(C) 1:1
(D) 2:1
(E) 3:1

GO ON TO NEXT PAGE ▷

CHAPTER 9: CHALLENGE QUESTIONS — Student-Produced Responses

1  2  3  4  5  6 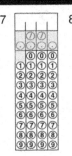 7  8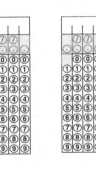

9.2 - 9.3

1. The longest diagonal of a single cube in a Rubik's Cube is $3\sqrt{8}$. There are 27 cubes that make up a full Rubik's Cube. What would the square of the longest diagonal be if two Rubik's Cubes were stacked on top of each other?

9.4

2. Suppose that a cylinder is stacked on top of another cylinder that has twice the first cylinder's radius. If the two cylinders have a total volume of 135π and they both share a height of 3, what is the radius of the larger cylinder?

3. A cylinder that is 5 inches tall has a volume of 125π. If this cylinder is inscribed in a sphere, what is the volume of the sphere? *(Round to the nearest whole number.)*

9.5

4. A hexagonal prism has a volume of 45 cubic inches. Suppose that a hexagonal pyramid is cut from the prism and the excess material is then molded into another hexagonal prism with the same base dimensions as the original. If a new hexagonal pyramid is then cut from this prism, what is this second pyramid's volume?

9.6

5. A rubber sphere with surface area 64π has a wooden cube inscribed within it, with all eight vertices touching the sphere's surface. What is the volume of the *rubber* in the object? *(Round to the nearest whole number.)*

9.8

6. A cone has a base radius of 3 inches and a height of 6 inches. If this cone is filled with water and the water is then poured into a sphere with a radius of 3 inches, what is the height of the water?

9.9

7. The surface area of a cylinder is 121 cm² and the surface area of a smaller cylinder is 81 cm². What is the result if the circumference of the larger cylinder's base is doubled, then divided by the circumference of the smaller cylinder's base?

8. Suppose that a right triangle is crossed by 10 parallel lines that break its height into 11 equal pieces. If every other section is shaded starting with the top piece, what is the ratio of the shaded to the unshaded areas?

GO ON TO NEXT PAGE ⇒

Multiple-Choice

1 Ⓐ Ⓑ Ⓒ Ⓓ Ⓔ
2 Ⓐ Ⓑ Ⓒ Ⓓ Ⓔ
3 Ⓐ Ⓑ Ⓒ Ⓓ Ⓔ
4 Ⓐ Ⓑ Ⓒ Ⓓ Ⓔ
Ⓐ Ⓑ Ⓒ Ⓓ Ⓔ
Ⓐ Ⓑ Ⓒ Ⓓ Ⓔ
Ⓐ Ⓑ Ⓒ Ⓓ Ⓔ
Ⓐ Ⓑ Ⓒ Ⓓ Ⓔ
Ⓐ Ⓑ Ⓒ Ⓓ Ⓔ
Ⓐ Ⓑ Ⓒ Ⓓ Ⓔ

CHAPTER
9
REVIEW

Student-Produced Responses

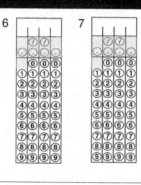  

5   6   7

1. What is the surface area of a rectangular prism that has a base with dimensions of 6 inches by 8 inches and a height of 3 inches?

(A) 72
(B) 96
(C) 108
(D) 180
(E) 192

2. Suppose that a cylinder with a radius of 5 and a height of 6 is carved out to make a cone with the same base and height. What is the volume of this cone?

(A) 20π
(B) 25π
(C) 40π
(D) 50π
(E) 75π

3. Which of the following steel rods is the largest that could be shipped in a shipping carton that is a cubic foot in size?

(A) 16 in.
(B) 17 in.
(C) 18 in.
(D) 19 in.
(E) 20 in.

4. Suppose that the radii of two circles are in a ratio of 1:3. What is the ratio of their circumferences?

(A) 1:3
(B) 1:9
(C) 1:27
(D) 1:81
(E) 1:243

5. Suppose that a sphere has a volume of 400π. If a cube is to be made that completely encloses the sphere, and the sphere touches the cube on all faces, what would the cube's volume be?

6. In the figure below, what is the ratio of the area of one of the smaller circles to the area of the larger circle?

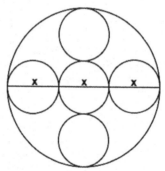

7. In the figure below, BAC is a semicircle in which arcs BA and AC have the same length, segments BX and YC have length 2, and XY has length 6. What is the area of triangle XAY?

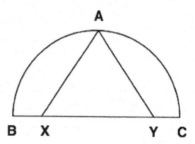

GO ON TO NEXT PAGE ⟹

Multiple-Choice

Student-Produced Responses

CHAPTER
1 - 9
CUMULATIVE
REVIEW

1 Ⓐ Ⓑ Ⓒ Ⓓ Ⓔ
2 Ⓐ Ⓑ Ⓒ Ⓓ Ⓔ
3 Ⓐ Ⓑ Ⓒ Ⓓ Ⓔ
4 Ⓐ Ⓑ Ⓒ Ⓓ Ⓔ
Ⓐ Ⓑ Ⓒ Ⓓ Ⓔ
Ⓐ Ⓑ Ⓒ Ⓓ Ⓔ
Ⓐ Ⓑ Ⓒ Ⓓ Ⓔ
Ⓐ Ⓑ Ⓒ Ⓓ Ⓔ
Ⓐ Ⓑ Ⓒ Ⓓ Ⓔ
Ⓐ Ⓑ Ⓒ Ⓓ Ⓔ

5  6  7  8

---

1. There are 75 adults and 1 child in a room. How many adults must leave the room so that 4 percent of the people remaining in the room are children?

(A) 56
(B) 51
(C) 50
(D) 48
(E) 46

2. In a list of 47 consecutive integers, the median is 31. What is the difference between the greatest integer and the least integers?

(A) 45
(B) 46
(C) 50
(D) 51
(E) 56

3. Kurt numbered the pages in his notebook from 1 to 66. In doing this, he wrote a digits. What is the value of a?

(A) 111
(B) 118
(C) 123
(D) 135
(E) 156

4. What is the slope of a line that is perpendicular to the line that goes through the point (4,6) and the origin?

(A) - 2/3
(B) - 4/5
(C) - 3/2
(D) 3/2
(E) 4/5

5. Eric's Bagel Shop sells sesame and plain bagels in whole wheat and multigrain varietes. Exactly 1/3 of the sesame bagels are whole wheat. If 2/5 of the plain bagels are multigrain and Eric sells 3 times as many plain bagels as sesame bagels, how many bagels of the 360 bagels sold on Monday are multigrain sesame bagels?

6. James has twice as many baseball cards as Chris does, and together they have half as many cards as Mary does. If the three collectors have 162 cards altogether, how many more baseball cards does Mary have than James?

7. Suppose that there are 424 students at a college orientation. If there are 124 more students from out of state than in state, how many in-state students are at the orientation?

8. Given the sequence below, what is the first term in the sequence if -512 is the 12th term?

... 64 , -128 , 256 , -512

STOP

Chapter 10: Miscellaneous Topics

6 Sections
78 Practice Questions

10.1 Venn Diagrams

10.1.1 **Sets**   **TOTAL = First Set + Second Set - Both + Neither**

Demonstration Examples

Demo 1: In a class of 52 students, 22 take Math 1 and 32 take Math 2. How many students take both Math 1 and Math 2?

Math 1 = 22   **Math 2 = 32**

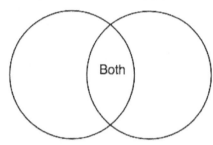

Both

TOTAL = Math 1 + Math 2 - Both + Neither
52   =   22   +   32   -   x   +   0

Solve --> x = 2

Demo 2: How many students are in Math 1 only?

Math 1 only is just Math 1 - Both. So, 22 - 2 = 20.

SAT Example and Technique Application:

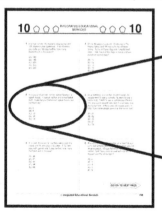

3. Suppose 32 students register for Spanish class, 42 register for French, and there is a total of 52 registrations. How many students are registered for French only?

(A) 24   **TOT = A + B - Both + Neither**
(B) 22   **52 = 32 + 42 - x + 0**
(C) 20
(D) 16   **So, x = 22. Therefore,**
(E) 10   **42 - 22 = 20 (French)**

10.1.1 **Let's Practice:**

1. Ⓐ Ⓑ Ⓒ Ⓓ Ⓔ   6. Ⓐ Ⓑ Ⓒ Ⓓ Ⓔ
2. Ⓐ Ⓑ Ⓒ Ⓓ Ⓔ   7. Ⓐ Ⓑ Ⓒ Ⓓ Ⓔ
3. Ⓐ Ⓑ Ⓒ Ⓓ Ⓔ   8. Ⓐ Ⓑ Ⓒ Ⓓ Ⓔ
4. Ⓐ Ⓑ Ⓒ Ⓓ Ⓔ   9. Ⓐ Ⓑ Ⓒ Ⓓ Ⓔ
5. Ⓐ Ⓑ Ⓒ Ⓓ Ⓔ   10. Ⓐ Ⓑ Ⓒ Ⓓ Ⓔ

1. In a high school, 90 students play soccer and 100 students play basketball. If 20 students play both and 30 play neither, how many students are in the school?

(A) 180
(B) 190
(C) 200
(D) 210
(E) 220

2. In a group of 25 men, 8 men speak German, 9 speak Italian, 11 speak neither and some speak both. How many of these men speak Italian, but not German?

(A) 6
(B) 8
(C) 10
(D) 12
(E) 13

3. In a club, there are 12 members who play only chess and 16 who play only poker. If 10 men play both games and 3 play neither, how many members are in the club?

(A) 38
(B) 41
(C) 42
(D) 44
(E) 48

4. Of the 30 dogs in a pound, 15 belong to the hound family and 18 belong to the retriever family. Some of these dogs are mixed-breed dogs. How many of the dogs in the pound are purebred hound dogs?

(A) 10
(B) 12
(C) 13
(D) 15
(E) 16

5. On a weekend at a certain movie theater, 24 people went to see a comedy, 30 went to see a horror flick, 7 went to see a completely different film, and some people saw both the comedy and the horror flick. If there were 50 movie goers in total, how many people saw only the horror flick?

(A) 27
(B) 25
(C) 22
(D) 21
(E) 19

6. At a zoo, a class of 30 children on a field trip split up to see different exhibits. If 14 children saw the snow leopards, 18 saw the white tiger, and 3 saw neither, how many children saw both the snow leopard and the white tiger?

(A) 2
(B) 3
(C) 4
(D) 5
(E) 6

GO ON TO NEXT PAGE ⟩

10.1.2 All/Some, Could/Must

Use Venn diagrams when you see the words "all" or "some".

Demonstration Venn Diagrams

ALL swimmers are runners:

Runners

Swimmers

SOME swimmers are runners:

Swimmers | Both | Runners

* When the question asks, "Which of the following <u>could</u> be true?," find one region of the diagram where the given condition <u>does</u> work.

* When the question asks, "Which of the following <u>must</u> be true?," find the region of the diagram where ithe given condition <u>does not</u> work.

<u>REMEMBER: Could --> Does , Must --> Doesn't</u>
This is how you should approach ANY question that involves Roman numerals and multiple correct answers even if it is not a logic question.

<u>SAT Example and Technique Application:</u>

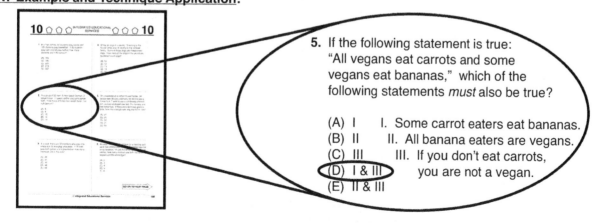

5. If the following statement is true: "All vegans eat carrots and some vegans eat bananas," which of the following statements *must* also be true?

(A) I
(B) II
(C) III
(D) I & III
(E) II & III

I. Some carrot eaters eat bananas.
II. All banana eaters are vegans.
III. If you don't eat carrots, you are not a vegan.

10.1.2 <u>Let's Practice:</u>

1. Ⓐ Ⓑ Ⓒ Ⓓ Ⓔ   6. Ⓐ Ⓑ Ⓒ Ⓓ Ⓔ
2. Ⓐ Ⓑ Ⓒ Ⓓ Ⓔ   7. Ⓐ Ⓑ Ⓒ Ⓓ Ⓔ
3. Ⓐ Ⓑ Ⓒ Ⓓ Ⓔ   8. Ⓐ Ⓑ Ⓒ Ⓓ Ⓔ
4. Ⓐ Ⓑ Ⓒ Ⓓ Ⓔ   9. Ⓐ Ⓑ Ⓒ Ⓓ Ⓔ
5. Ⓐ Ⓑ Ⓒ Ⓓ Ⓔ   10. Ⓐ Ⓑ Ⓒ Ⓓ Ⓔ

1. If the following statement is true: "Some skiers are bikers and no biker is a golfer," which of the following *could* also be true?

    I.   Some golfers do not ski
    II. All skiers are golfers
    III. No golfer is a skier

(A) I only
(B) II only
(C) III only
(D) I and III
(E) None

2. If the following statement is true, "All IES students are intelligent and some IES students score a 2400 on the SAT," which of the following statements *must* also be true?

    I.   If you did not score a 2400, you are not an IES student.
    II. If you are intelligent, you will score a 2400.
    III. If you score a 2400, you are intelligent.

(A) I only
(B) II only
(C) III only
(D) I, II, and III
(E) None

3. A box of blocks contains blocks of different shapes and colors. If all the star-shaped blocks are blue, which of the following *must* be true?

    I.   If a block is not a star, it is not blue.
    II. If a block is blue then it is a star.
    III. If a block is not blue, then it is not a star.

(A) I only
(B) II only
(C) III only
(D) I and II
(E) All of the above

4. All red fish in a pond are predators and all predatory fish are prey for birds. Which of the following *must* be true?

    I.   Fish that are prey for birds are also predators.
    II. Red fish are prey for birds.
    III. If a fish is not prey for a bird, it is not a predatory fish.

(A) I only
(B) I and II
(C) II and III
(D) I, II, and III
(E) None

5. All birds-of-paradise have long tail feathers and some birds-of-paradise have bright feathers. Which of the following *could* be true?

    I.   If a bird has long tail feathers, it is a bird-of-paradise
    II. A bird that is not a bird-of-paradise does not have long feathers.
    III. A brightly-feathered bird that also has long tail feathers is a bird of paradise.

(A) I, II, and III
(B) I and II
(C) II and III
(D) I only
(E) None

GO ON TO NEXT PAGE ⟩

10.1.3 **Three-Circle Venn Diagrams**

When looking at sets, you will sometimes see a Venn diagram with multiple overlaps. Use your understanding of visual cues to answer the following questions:

Demonstration Example

Demo: According to the Venn diagram below, how many people either eat cantaloupe or eat both apples and bananas?

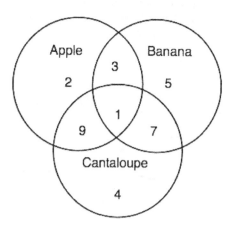

Here we will be counting for all the regions that are in both Apple and Banana (3 + 1) and adding those regions to all of the other regions that are within Cantaloupe (9 + 7 + 4). Be sure to count each region only once, even though the middle follows both of the question's conditions. The answer is 24.

10.1.3 **Let's Practice:** (Use The Diagram from the demonstration above to answer the following.)

1. Ⓐ Ⓑ Ⓒ Ⓓ Ⓔ   6. Ⓐ Ⓑ Ⓒ Ⓓ Ⓔ
2. Ⓐ Ⓑ Ⓒ Ⓓ Ⓔ   7. Ⓐ Ⓑ Ⓒ Ⓓ Ⓔ
3. Ⓐ Ⓑ Ⓒ Ⓓ Ⓔ   8. Ⓐ Ⓑ Ⓒ Ⓓ Ⓔ
4. Ⓐ Ⓑ Ⓒ Ⓓ Ⓔ   9. Ⓐ Ⓑ Ⓒ Ⓓ Ⓔ
5. Ⓐ Ⓑ Ⓒ Ⓓ Ⓔ   10. Ⓐ Ⓑ Ⓒ Ⓓ Ⓔ

1. How many people eat both bananas and cantaloupes?

(A) 1
(B) 7
(C) 8
(D) 20
(E) 29

2. How many people eat exactly two of the three fruits?

(A) 3
(B) 7
(C) 9
(D) 19
(E) 20

3. How many people eat apples or bananas, but not both?

(A) 7
(B) 10
(C) 23
(D) 26
(E) 27

GO ON TO NEXT PAGE

10.2 Matrices

When solving problems with multiple variables, you should use a matrix and fill in the cells as you arrive at new information. Remember to *Aim for Your Target* and circle the cell you need.

Demonstration Example

Demo: In a group of 100 men and women, some individuals are married and some are single. There are 20 more women than men and there are 40 more single individuals than married individuals. There are also 12 married men. How many single women are there?

	SINGLE	MARRIED	TOTAL
MEN		12	
WOMEN	⬭		
TOTAL			100

After you have filled in the information that you immediately know, use *Split the Difference* to find the numbers for men/women and single/married. Then fill in the cells you know and add or subtract to reach your target.

	SINGLE	MARRIED	TOTAL
MEN	28	12	40
WOMEN	42		60
TOTAL	70	30	100

10.2 Let's Practice:

1. Ⓐ Ⓑ Ⓒ Ⓓ Ⓔ
2. Ⓐ Ⓑ Ⓒ Ⓓ Ⓔ
3. Ⓐ Ⓑ Ⓒ Ⓓ Ⓔ
4. Ⓐ Ⓑ Ⓒ Ⓓ Ⓔ
5. Ⓐ Ⓑ Ⓒ Ⓓ Ⓔ

6. Ⓐ Ⓑ Ⓒ Ⓓ Ⓔ
7. Ⓐ Ⓑ Ⓒ Ⓓ Ⓔ
8. Ⓐ Ⓑ Ⓒ Ⓓ Ⓔ
9. Ⓐ Ⓑ Ⓒ Ⓓ Ⓔ
10. Ⓐ Ⓑ Ⓒ Ⓓ Ⓔ

1. At the Medical Institute of Dahomey, 80% of the graduates are doctors and the rest are nurses. 60% of the graduates are female. If 25% of the female graduates are nurses, what percent of all graduates are male nurses?

(A) 5
(B) 10
(C) 15
(D) 20
(E) 25

GO ON TO NEXT PAGE ⟩

2. At Bill's Used Cars there are 2-door and 4-door cars which come in either red or green. 20% of the green cars are 4-door, and 50% of the red cars are 2-door. If there are 70 red cars and 30 green cars, what is the ratio of 2-door to 4-door cars at Bill's Used Cars?

(A) 3/2
(B) 4/9
(C) 5/7
(D) 59/41
(E) 39/37

3. In a novelty book store, the books have been separated and shelved either as works by first-time authors or by more established authors. These sections have been further divided depending on whether the book is a novel or an anthology of short stories. 48 percent of all the books in this store are novels and 58 percent of all the books are by first-time authors. If 8 percent of all the books in the store are novels by first-time novelists, what percent of all books are short story collections by experienced authors?

(A) 2
(B) 4
(C) 8
(D) 10
(E) 12

10.3 Contrapositives (If/Then)

A contrapositive is the exact, logical, opposite statement. When you see the words "if" or "then" in a sentence, like the sentence in the demonstration example below, do NOT look at the answer choices to try and figure out which one makes sense in your head. Instead, write the contrapositive as demonstrated and find it among the answer choices.

Demonstration Example

Demo: If it is raining, then the grass is wet. Which of the following must also be true?

I. If it is not raining, the grass is not wet.
II. If the grass is wet, it is raining.
III. If the grass is not wet, it is not raining.

If <u>it is raining</u> .... then <u>the grass is wet</u>.

If <u>the grass is NOT wet</u> .... then <u>it is NOT raining</u>.

Step 1: Underline the parts after "if" and "then."
Step 2: Switch the sentence fragments and make them into exact opposites.

Therefore, the correct answer is <u>III</u>.

It is easy to fall into the trap of reading through the answer choices and selecting the wrong one because it sounds right when you read it to yourself. This is both time-consuming and inaccurate. You cannot bring any real-world knowledge or experience into these problems. Do not assume anything. Just apply the technique above, find the correct answer and move on. It's that simple.

10.3 Let's Practice:

1. Ⓐ Ⓑ Ⓒ Ⓓ Ⓔ   6. Ⓐ Ⓑ Ⓒ Ⓓ Ⓔ
2. Ⓐ Ⓑ Ⓒ Ⓓ Ⓔ   7. Ⓐ Ⓑ Ⓒ Ⓓ Ⓔ
3. Ⓐ Ⓑ Ⓒ Ⓓ Ⓔ   8. Ⓐ Ⓑ Ⓒ Ⓓ Ⓔ
4. Ⓐ Ⓑ Ⓒ Ⓓ Ⓔ   9. Ⓐ Ⓑ Ⓒ Ⓓ Ⓔ
5. Ⓐ Ⓑ Ⓒ Ⓓ Ⓔ   10. Ⓐ Ⓑ Ⓒ Ⓓ Ⓔ

1. If I am smiling, I am not sad.
Which of the following is logically true?

   I. If I am not sad, I am smiling.
   II. If I am sad, I am not smiling.
   III. If I am not smiling, I am sad.

   (A) I only
   (B) II only
   (C) III only
   (D) I and III
   (E) None

2. If it is bright outside, then the sun is shining.
Which of the following is logically true?

   I. If the sun is not shining, then it is not bright outside.
   II. If the sun is shining, then it is not bright outside.
   III. If it is not bright outside, then the sun is not shining.

   (A) I only
   (B) II only
   (C) III only
   (D) I and II
   (E) None

3. If a man is evil, then he is greedy.
Which of the following is logically true?

   I. If a man is greedy, then he is evil.
   II. If a man is not greedy, then he is not evil
   III. If a man is not evil, then he is not greedy.

   (A) I only
   (B) II only
   (C) III only
   (D) I and II
   (E) II and III

4. If something is on fire, then it is smoking.
Which of the following is logically true?

   I. If something is smoking, then it is on fire.
   II. If something is not on fire, then it is not smoking.
   III. If something is not smoking, then it is not on fire.

   (A) I only
   (B) II only
   (C) III only
   (D) I and III
   (E) None

5. If X is greater than or equal to 10, then Y is less than 3. Which of the following is logically true?

   I. If Y is greater than or equal to 3, then X is less than 10.
   II. If X is less than 10, then Y is greater than or equal to 3.
   III. If Y is less than 3, then X is greater than or equal to 10.

   (A) I only
   (B) II only
   (C) III only
   (D) I and II
   (E) None

6. If something is not dense, then it is light.
Which of the following is logically true?

   I. If something is light, then it is not dense.
   II. If something is heavy, then it is dense.
   III. If something is dense, then it is heavy.

   (A) I only
   (B) II only
   (C) III only
   (D) I and III
   (E) II and III

GO ON TO NEXT PAGE ⟩

10   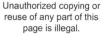   10

Unauthorized copying or reuse of any part of this page is illegal.

10.4 Combinations vs. Permutations

10.4.1 Basic Combinations and Permutations

Combinations are unique "groups" that can be made from a set of elements. The order in which you choose things is NOT important. Ex: When choosing a pair out of 3 players A, B, and C, picking players A and B is the same as if you were to pick players B and A. The order in which these players are selected does not make any difference.

Permutations are unique "arrangements" of these elements. You may have the same group of elements, but you can arrange them differently. So, in this case, order DOES matter. For example, 3, 2, and 1 are the same digits, but 321 and 123 are two different numbers. (Hint: the words "order" or "arrange" are often good clues that you are looking at a permutation question.)

You will have to decide for yourself whether a question is asking you to <u>group</u> things or <u>arrange</u> them.

What are you trying to do?			
Make Groups...		Make Arrangements...	
From the same set:	**From different sets:**	**Without restrictions:**	**With restrictions:**
$_nC_r$ How many ways can you make groups of "r" things out of the "n" things you have to choose from?	Matching Multiply all of your singular $_nC_r$'s together.	$_nP_r$ How many different ways can you arrange "r" things out of the "n" things you have to choose from? Hint: If you are trying to arrange all "n" things, the answer is (n!).	P(R!)(U!) Use the equation above for objects. Otherwise use dashes for restrictions on numbers.

$_nC_r$ and $_nP_r$

Use these to find the number of different subsets of "r" elements that can be drawn from a main set of "n" elements. When the order of the "r" elements in the subset matters, use $_nP_r$, not $_nC_r$.

$_nC_r$ - **There are $_5C_3$ = 10 groups of 3 children that can be made from a class of 5 children.**

$_nP_r$ - **There are $_6P_3$ = 120 3-digit integers that can be made from the digits: 1 2 3 4 5 6**

10.4.1 **Let's Practice:**

1. Ⓐ Ⓑ Ⓒ Ⓓ Ⓔ   6. Ⓐ Ⓑ Ⓒ Ⓓ Ⓔ
2. Ⓐ Ⓑ Ⓒ Ⓓ Ⓔ   7. Ⓐ Ⓑ Ⓒ Ⓓ Ⓔ
3. Ⓐ Ⓑ Ⓒ Ⓓ Ⓔ   8. Ⓐ Ⓑ Ⓒ Ⓓ Ⓔ
4. Ⓐ Ⓑ Ⓒ Ⓓ Ⓔ   9. Ⓐ Ⓑ Ⓒ Ⓓ Ⓔ
5. Ⓐ Ⓑ Ⓒ Ⓓ Ⓔ   10. Ⓐ Ⓑ Ⓒ Ⓓ Ⓔ

1. Billy has six books, but room on his shelf for only four books. How many ways can he arrange his books?

   (A)   15
   (B)   36
   (C)   120
   (D)   360
   (E)   400

2. Roger is bringing his four young cousins to the movies. How many arrangements can he make to seat his cousins in the theater?

   (A)   1
   (B)   4
   (C)   24
   (D)   36
   (E)   48

3. What is the total number of different 4-digit positive integers that can be created using the following digits, with no digit repeated within the integer: 4 5 6 7 8 9

   (A)   15
   (B)   36
   (C)   120
   (D)   240
   (E)   360

4. How many direct paths can there be linking every 2 cities in a state with a total of 6 cities?

   (A)   10
   (B)   12
   (C)   15
   (D)   24
   (E)   30

5. There are ten people in a flag football lineup. How many ways are there to select a team of 4 players?

   (A)   180
   (B)   210
   (C)   420
   (D)   3200
   (E)   5040

6. Find the total number of different 5-digit positive integers that can be generated using the following digits, where no digit is repeated within a given integer: 5 6 7 8 9.

   (A)   1
   (B)   120
   (C)   150
   (D)   180
   (E)   2100

7. Find the total number of different 3-digit positive integers that can be generated using the following digits, where no digit is repeated within an integer and the units digit is 9: 3 4 5 6 7 8 9.

   (A)   15
   (B)   30
   (C)   50
   (D)   60
   (E)   80

8. How many gifts are exchanged in a family of 5 if every member gives one gift to each other family member?

   (A)   5
   (B)   10
   (C)   15
   (D)   16
   (E)   20

GO ON TO NEXT PAGE ➔

10.4.2 **Matching (Complex Combinations)**

Multiply together when you are combining elements from multiple sets. This idea also known as the FCP, the Fundamental Counting Principle: If there are A ways to do one thing and B ways to do another, the number of ways to do them together is A x B. This applies when we are talking about more than 2 things as well.

Demonstration Examples

Demo 1: A gift basket requires 2 kinds of potpourri and 1 kind of scented candle. If there are 6 kinds of potpourri and 3 kinds of candles to choose from, how many gift basket combinations can be made?

$_6C_2$ = **15 ways to pick the 2 kinds of potpourri**
$_3C_1$ = **3 ways to pick the scented candle**

So, **(15)(3) = 45 gift basket combinations.**

Demo 2: For a car, there are 5 choices for the paint color and 3 choices for the tire size. How many paint/tire combinations can be made?

$_5C_1$ = **5 ways to pick the paint**
$_3C_1$ = **3 ways to pick the tire size**

So, **(5)(3) = 15 car styles to choose from.**

SAT Example and Technique Application:

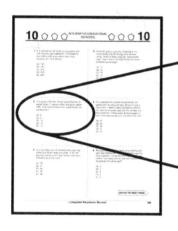

8. In a certain firm, a consulting team consists of 1 analyst and 2 accountants. If there are 4 analysts and 6 accountants to choose from, how many teams can be made at this firm?

(A) 60
(B) 48 $_4C_1$ x $_6C_2$ =
(C) 30   4 x 15 = **60**
(D) 20
(E) 15

10.4.2 **Let's Practice:**

1. Ⓐ Ⓑ Ⓒ Ⓓ Ⓔ   6. Ⓐ Ⓑ Ⓒ Ⓓ Ⓔ
2. Ⓐ Ⓑ Ⓒ Ⓓ Ⓔ   7. Ⓐ Ⓑ Ⓒ Ⓓ Ⓔ
3. Ⓐ Ⓑ Ⓒ Ⓓ Ⓔ   8. Ⓐ Ⓑ Ⓒ Ⓓ Ⓔ
4. Ⓐ Ⓑ Ⓒ Ⓓ Ⓔ   9. Ⓐ Ⓑ Ⓒ Ⓓ Ⓔ
5. Ⓐ Ⓑ Ⓒ Ⓓ Ⓔ   10. Ⓐ Ⓑ Ⓒ Ⓓ Ⓔ

1. Melanie is planning her outfits for the week. If she has 3 pairs of jeans, 4 vests, and 6 shirts, how many outfits can she make if she wears one pair of jeans, one shirt, and one vest each day?

   (A)   60
   (B)   72
   (C)   75
   (D)   80
   (E)   100

2. Given the data from the previous question, how many outfits can Melanie make if she layers two shirts and one vest for a more fashionable look?

   (A)   180
   (B)   240
   (C)   320
   (D)   360
   (E)   420

3. Monica has 36 potential outfits which she can make by combining a shirt, a pair of pants, and a pair of shoes. Which of the following *cannot* be the number of shirts Monica has?

   (A)   2
   (B)   3
   (C)   4
   (D)   5
   (E)   6

10.4.3 **Restricted Arrangements (Complex Permutations)**

Remember to use (P)(R!)(U!), where P is the number of possible ways to meet a restriction, R is the number of restricted objects, and U is the number of unrestricted objects.

Demonstration Example

> **Demo:** Damien has 5 trophies: a soccer trophy, a hockey trophy, a baseball trophy, a karate trophy, and a basketball trophy. If he wants to arrange his trophies on his mantle in such a way that the karate trophy is never at either end, how many ways can he arrange his trophies?
>
> ____ ____ ____ ____ ____
>
> **In this problem, we face a restriction: the karate trophy cannot be on either end. So, this gives us the following:**
>
> **P = 3, There are three locations where the karate trophy CAN be located.**
> **R = 1, There is one object, the karate trophy, that has a restriction on it.**
> **U = 4, There are 4 objects that have no restrictions on them.**
>
> **So, (P)(R!)(U!) = (3)(1!)(4!) = 72 possible arrangements.**

SAT Example and Technique Application:

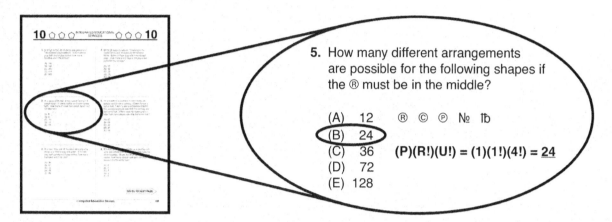

5. How many different arrangements are possible for the following shapes if the ® must be in the middle?

(A) 12
(B) 24
(C) 36
(D) 72
(E) 128

$(P)(R!)(U!) = (1)(1!)(4!) = \underline{24}$

Let's Practice:

1. Ⓐ Ⓑ Ⓒ Ⓓ Ⓔ   6. Ⓐ Ⓑ Ⓒ Ⓓ Ⓔ
2. Ⓐ Ⓑ Ⓒ Ⓓ Ⓔ   7. Ⓐ Ⓑ Ⓒ Ⓓ Ⓔ
3. Ⓐ Ⓑ Ⓒ Ⓓ Ⓔ   8. Ⓐ Ⓑ Ⓒ Ⓓ Ⓔ
4. Ⓐ Ⓑ Ⓒ Ⓓ Ⓔ   9. Ⓐ Ⓑ Ⓒ Ⓓ Ⓔ
5. Ⓐ Ⓑ Ⓒ Ⓓ Ⓔ   10. Ⓐ Ⓑ Ⓒ Ⓓ Ⓔ

1. How many different arrangements are possible for the following shapes:

if the ♣ can never be on either end?

(A) 320
(B) 360
(C) 480
(D) 500
(E) 520

2. How many ways can you arrange the following shapes if the ♡ cannot be in the center?

(A) 96
(B) 192
(C) 240
(D) 288
(E) 360

3. Five cards are drawn at random from a deck of cards. These cards are an Ace, a King, a Queen, a Jack, and a Joker. How many ways can the cards be arranged in a row if the Ace must be in the middle and the King cannot be at either end?

(A) 10
(B) 12
(C) 18
(D) 24
(E) 36

4. Simon wants to arrange seven pictures of his friends and family on a mantle. How many ways can he arrange the pictures if the picture of his girlfriend must remain in the center?

(A) 24
(B) 72
(C) 96
(D) 240
(E) 720

GO ON TO NEXT PAGE ⟩

10.4 **Combination / Permutation Mixed Practice**

1   2   3   4   5   6   7   8

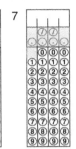
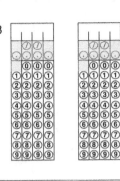

1. Adam is looking at the stars in the sky and sees a constellation made up of 7 stars. He starts tracing imaginary lines between pairs of stars with his finger. How many such lines can he draw?

2. A superhero has 3 capes (red, white, and blue), 3 bullet-proof vests (red, white, and blue), and 3 pairs of tights (red, white, and blue). He wants to design an outfit consisting of a cape, a vest, and tights where each item is a different color. How many such outfits can he make?

3. An art connoisseur is hanging paintings on a wall. He bought 6 paintings at Sotheby's, but only has room on his wall for three paintings. How many ways can he arrange the paintings on his wall?

4. A world bank is making its employees create PIN ID's in order to access client accounts. A password must contain 4 characters, the first 3 being different digits less than 8 and the last being a letter from the alphabet. How many such passwords can be created?

5. A gardener is planting five rows of trees, each row containing a different type of tree. He is planting peach trees, cherry trees, apple trees, pear trees, and plum trees. How many ways can he arrange the rows of trees if the row containing peach trees cannot be the first or last row?

6. A customer at a certain shop can make a custom fruit smoothie by choosing 3 different fruit flavors. The store offers 7 fruit flavors to choose from. How many possible smoothies can be made?

7. A rare book collector has obtained first edition copies of 8 different novels, but only has room on his shelf for 5. How many ways can he arrange the 5 books?

8. The code to a certain safe must be a 4-digit number that is greater than 7000, and each digit can only be used once. How many possible codes are there?

GO ON TO NEXT PAGE ▷

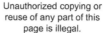

10.5 Probability

The **probability** that a particular event will occur is the number of favorable or desired outcomes, divided by the total number of possible outcomes. If there are 10 dogs in a pound and 2 of them are greyhounds, the probability of picking a greyhound at random is 2/10 or 20%.

Favorable Outcomes

Total Outcomes

Independent Events are defined as outcomes that have no effect on one another. For example, if I flip a coin and get heads, a common mistake is to think that since the odds are 50:50, a second flip will likely give me tails. This is false. One toss of a coin has nothing to do with another toss: the odds are still 50:50. I am still just as likely to get heads as tails. If you are looking for the probability of multiple independent events occurring, you just multiply together the single probabilities of the individual events.

Examples of the probability of singular or multiple independent events:

A. What is the probability of getting a number less than 3 when you roll a die?
There are 2 numbers less than 3. So, 2/6, which is a probability of 1/3.

B. What is the probability of getting an odd number when you roll a die?
There are 3 odd numbers on the die. So, 3/6, which is a probability of 1/2.

C. You roll the die 2 times, what is the probability of getting a number less than 3 the first time and an odd number the second time?
Multiply the two probabilities: (1/3) x (1/2) = 1/6.

The sum of all probabilities of an event is 1 and the probability that an event does NOT occur is 1 minus the probability that the event WILL occur.

Demonstration Example

> **Demo:** Sarah has 6 matching pairs of winter coats and cashmere sweaters. If Sarah picks a winter coat and a sweater at random, what is the probability that they will NOT match?
>
> **What we can do is find the probability that she chooses a matching pair and then subtract that from 1.**
>
> **There are 6 x 6 = 36 total ways she can choose the pair, and there are 6 matching pairs. So, the probability of choosing a matching pair is 6/36 or 1/6.**
>
> **Then: 1 - 1/6 = 5/6   It's that simple.**

SAT Example and Technique Application:

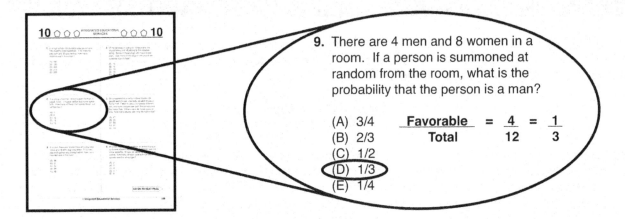

9. There are 4 men and 8 women in a room. If a person is summoned at random from the room, what is the probability that the person is a man?

(A) 3/4
(B) 2/3
(C) 1/2
(D) 1/3
(E) 1/4

$$\frac{\text{Favorable}}{\text{Total}} = \frac{4}{12} = \frac{1}{3}$$

10.5 **Let's Practice:**

1. Ⓐ Ⓑ Ⓒ Ⓓ Ⓔ   6. Ⓐ Ⓑ Ⓒ Ⓓ Ⓔ
2. Ⓐ Ⓑ Ⓒ Ⓓ Ⓔ   7. Ⓐ Ⓑ Ⓒ Ⓓ Ⓔ
3. Ⓐ Ⓑ Ⓒ Ⓓ Ⓔ   8. Ⓐ Ⓑ Ⓒ Ⓓ Ⓔ
4. Ⓐ Ⓑ Ⓒ Ⓓ Ⓔ   9. Ⓐ Ⓑ Ⓒ Ⓓ Ⓔ
5. Ⓐ Ⓑ Ⓒ Ⓓ Ⓔ   10. Ⓐ Ⓑ Ⓒ Ⓓ Ⓔ

1. The probability that Carla will go to the gym is 1/5. The probability that Kristi will go to the gym is 1/4. The probability that Karl will NOT go to the gym is 1/3. What is the probability that the 2 women will not go to the gym, but that Karl will?

(A) 1/60
(B) 1/20
(C) 1/12
(D) 2/5
(E) 3/5

2. There are ten cats at a shelter and each cat is the sibling of exactly one other cat. If a little girl buys 2 cats at random from the shelter, what is the probability that the cats are NOT siblings?

(A) 9/10
(B) 8/9
(C) 2/3
(D) 1/3
(E) 1/9

3. A coin is tossed 5 successive times. What is the probability of getting exactly 4 heads and one tails?

(A) 5/64
(B) 5/32
(C) 5/16
(D) 5/8
(E) 8/5

4. A pair of die is cast. What is the probability that the sum of the rolled numbers is greater than 8?

(A) 5/6
(B) 5/12
(C) 5/18
(D) 7/20
(E) 7/24

5. What is the probability that you will roll a single die three times in a row so that you get a number greater than 5 the first time, a number less than 4 the second time, and a prime number the third time?

(A) 1/24
(B) 1/18
(C) 1/10
(D) 1/5
(E) 1/2

GO ON TO NEXT PAGE >

10.6 Counting Problems

The Golden Rule of Counting:

Last - First + 1 (Inclusive)   Last - First - 1 (Exclusive)

Demonstration Example

Demo: If you read a book from page 22 to page 54, how many pages have you read?

Inclusive, since you have read pages 22 and 54...

54 - 22 + 1 = **33 Pages**

10.6 **Let's Practice:**

1. Ⓐ Ⓑ Ⓒ Ⓓ Ⓔ   6. Ⓐ Ⓑ Ⓒ Ⓓ Ⓔ
2. Ⓐ Ⓑ Ⓒ Ⓓ Ⓔ   7. Ⓐ Ⓑ Ⓒ Ⓓ Ⓔ
3. Ⓐ Ⓑ Ⓒ Ⓓ Ⓔ   8. Ⓐ Ⓑ Ⓒ Ⓓ Ⓔ
4. Ⓐ Ⓑ Ⓒ Ⓓ Ⓔ   9. Ⓐ Ⓑ Ⓒ Ⓓ Ⓔ
5. Ⓐ Ⓑ Ⓒ Ⓓ Ⓔ   10. Ⓐ Ⓑ Ⓒ Ⓓ Ⓔ

1. How many integers are there from 11 to 100, inclusive?

(A) 85
(B) 88
(C) 89
(D) 90
(E) 92

2. How many integers are there from -10 to 20, inclusive?

(A) 9
(B) 11
(C) 29
(D) 30
(E) 31

3. The number of integers that occur from -50 to x, exclusive, is 49. What is the value of x?

(A) 51
(B) 50
(C) 49
(D) 1
(E) 0

4. Two sides of a triangle are 100 and 108. How many integer solutions are there for the third side of the triangle?

(A) 7
(B) 8
(C) 9
(D) 199
(E) 201

5. A ten-day ritual begins on the 5th of February. On what day does the ritual end?

(A) 14th
(B) 15th
(C) 16th
(D) 18th
(E) 24th

6. A certain exam lasts 3 hours and 45 minutes, including three breaks of 15 minutes each. How many minutes does each segment of the exam last?

(A) 30
(B) 35
(C) 40
(D) 45
(E) 60

GO ON TO NEXT PAGE ➡

CHAPTER 10: CHALLENGE QUESTIONS

Student-Produced Responses

1  2 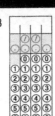 3  4  5  6  7

10.1

1. Suppose there are 22 students who take just Algebra 1 and that the number of students who take both Algebra 1 and Geometry is equal to the number of students who take just Geometry. If there are 24 students who take just one class, how many students take both Algebra 1 and Geometry?

2. If the following statement is true: "All swimmers eat grapefruit, all cyclists eat grapefruit, all kayakers eat grapefruit, and some people swim, cycle, and kayak," which of the following statements <u>must</u> be true?

   I.   Some cyclist ride kayaks as well.
   II. Some people swim and kayak only.
   III. Some people cycle only.

   (Bubble the corresponding number.)

   1. I only   4. I and II
   2. II only   5. I, II, and III
   3. III only   6. None of the above

10.4

3. Suppose that I am trying to select the following supplies for school: pens, rulers, and notepads. There are 3 stores in our town with unique items and each store has 3 different types of pens, 3 different types of rulers, and 2 different types of notepads. How many combinations of pen, ruler, and notepad can I make if I am buying 2 different pens, 2 different rulers and 1 notepad?

4. How many ways can you arrange the following objects if the ☂ and the ♣ cannot be on either end, or in the middle position?

   ☂   ♣   ★   ♠   ♣   ♥   ♦

10.5

5. What is the probability of selecting a king followed by another king given the following conditions: at first, the deck only has face cards and after you select the first king, the rest of the cards are returned to the deck before you choose the second king.

6. What is the probability of being dealt a full house, which is a hand of cards that consists of 3 of a kind and two of a kind, given that you already have 3 queens and you are the only person to whom cards are being dealt?

10.6

7. Suppose you have the following assignment:

   "Read pages 21-37, 35-51, 58-59, and pages 68-83, but not the first half of page 68 and the last half of page 83."

   How many pages will you read in total?

GO ON TO NEXT PAGE ⟩

Multiple-Choice	Student-Produced Responses

CHAPTER 10 REVIEW

1 Ⓐ Ⓑ Ⓒ Ⓓ Ⓔ
2 Ⓐ Ⓑ Ⓒ Ⓓ Ⓔ
3 Ⓐ Ⓑ Ⓒ Ⓓ Ⓔ
Ⓐ Ⓑ Ⓒ Ⓓ Ⓔ
Ⓐ Ⓑ Ⓒ Ⓓ Ⓔ
Ⓐ Ⓑ Ⓒ Ⓓ Ⓔ
Ⓐ Ⓑ Ⓒ Ⓓ Ⓔ
Ⓐ Ⓑ Ⓒ Ⓓ Ⓔ
Ⓐ Ⓑ Ⓒ Ⓓ Ⓔ
Ⓐ Ⓑ Ⓒ Ⓓ Ⓔ

4  5  6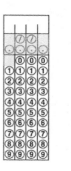

1. Three dice are rolled and the numbers are added up. What is the probability that the sum of the numbers rolled is equal to 16?

   (A) 1/36
   (B) 1/24
   (C) 1/12
   (D) 1/8
   (E) 1/4

2. If 25% of the students in Metuchen High School are soccer players, 65% of the soccer players are male, and there are 21 female soccer players, how many students are in the school?

   (A) 120
   (B) 240
   (C) 320
   (D) 360
   (E) 480

3. Set A has a members and set B has b members. Set C consists of all members that are either in set A or Set B with the exception of z common members, where $z > 0$. What algebraic expression represents the number of members in Set C?

   (A) $a + b - z$
   (B) $a + b - 2z$
   (C) $a + b + z$
   (D) $a + b + 2z$
   (E) $a - b + z$

4. In the figure below, circular region A represents all integers from 20 to 100 (inclusive), circular region B represents all integers that are multiples of 4, and circular region C represents all squares of integers. How many numbers are represented by the region marked with an X?

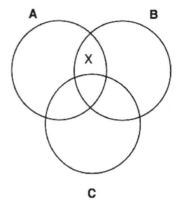

5. Suppose there are 6 friends going to the movies:

   Jill, Janice, Joy, James, Jon, and Jimmy.

   How many seating arrangements are there if James sits next to Jill on Jill's right, Jon sits next to Janice on Janice's right, and Jimmy sits next to Joy on Joy's right?

6. Challenge: Given the group of friends from question number 5, how many seating arrangements are there if the men must sit next to the same women as before, but can sit on either side?

GO ON TO NEXT PAGE ⟩

Multiple-Choice

Student-Produced Responses

CHAPTER 1 - 10 CUMULATIVE REVIEW

1 Ⓐ Ⓑ Ⓒ Ⓓ Ⓔ
2 Ⓐ Ⓑ Ⓒ Ⓓ Ⓔ
3 Ⓐ Ⓑ Ⓒ Ⓓ Ⓔ
4 Ⓐ Ⓑ Ⓒ Ⓓ Ⓔ
Ⓐ Ⓑ Ⓒ Ⓓ Ⓔ
Ⓐ Ⓑ Ⓒ Ⓓ Ⓔ
Ⓐ Ⓑ Ⓒ Ⓓ Ⓔ
Ⓐ Ⓑ Ⓒ Ⓓ Ⓔ
Ⓐ Ⓑ Ⓒ Ⓓ Ⓔ
Ⓐ Ⓑ Ⓒ Ⓓ Ⓔ

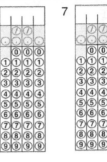
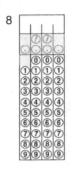

1. Three dice are rolled and the numbers are added up. What is the probability that the sum of the numbers rolled is equal to 15?

(A) 1/216
(B) 3/216
(C) 3/108
(D) 5/108
(E) 1/36

2. In a list of 91 consecutive integers, the median is 57. What is the greatest integer?

(A) 95
(B) 98
(C) 99
(D) 101
(E) 102

3. $x^2 - 9x + mx + m = (x - 2)^2 + 1$. What is m?

(A) 3
(B) 5
(C) 6
(D) 7
(E) 9

4. The sequence AABBBCCCAABBBCCC... is repeated. How many "B" terms are there in the first 500 terms of the sequence?

(A) 90
(B) 94
(C) 180
(D) 188
(E) 192

5. The value of the Browns' beach house increased by 30% in the first year the family owned it, so the Browns decided to sell. Unfortunately, before they sold, a hurricane damaged the area, devaluing the house by 25%. To apply for a government loan for first-year homes that lose value, a family needs to have total losses of 5% or greater. The Browns applied for the loan and unfortunately were not approved because the losses were only at what percent?

6. Erin has been given $85 for her birthday present. She really wants a new jacket. The nearest department store sells jackets for the prices in the table below. The store is having a 20% off sale and taxes are 7%. If Erin has a 15% off coupon, what is the most expensive jacket she can buy in light of the prices in the chart below? *(Bubble the corresponding number.)*

Jacket	Price	
Cotton	$105	(1)
Fleece	$110	(2)
Trench	$115	(3)
Suede	$118	(4)
Leather	$120	(5)

7. The average of m and m - 4 is j. The average of 8 and -2m is k. What is the average of j and k?

8. Suppose that two perpendicular lines cross at the point (3,3). If the y-intercept of one of the lines is (0,7), what is the y-intercept of the other line?

STOP

Practice Test A

SECTION 1
Time - 25 Minutes
20 Questions

Answer the questions in this section directly on the test or use a grid-in answer sheet.

Directions: For this section, solve each problem and decide which is the best of the multiple-choice answers. Fill in the corresponding circle in the answer bank or on a separate answer sheet. You may use the space available for scratch work.

Reference Information

$A = \pi r^2$
$C = 2\pi r$ $A = lw$ $A = 1/2bh$ $V = lwh$ $V = \pi r^2 h$ $c^2 = a^2 + b^2$ Special Right Triangles

The total number of degrees of arc in a circle is 360.
The sum of the measures of the three angles of a triangle is 180 degrees.

---

1. Set *M* consists of all numbers that are prime or multiples of 5. Which of the following numbers is in set *M*?

(A) 2
(B) 6
(C) 9
(D) 18
(E) 39

2. If 2.5x - 1 is divided by 7, the result has an integer value. Which of the following could be a value of x?

(A) 4
(B) 8
(C) 10
(D) 12
(E) 20

GO ON TO NEXT PAGE ⟩

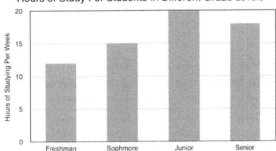

Hours of Study For Students In Different Grade Levels

3. The bar graph above shows the hours of studying per week completed by students at different grade levels. Students study what percent more per week in their junior year than in their sophomore year?

(A) 20
(B) 25
(C) 30
(D) 33
(E) 50

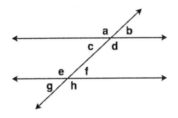

4. In the figure above, how many angles are supplementary to ∠g?

(A) 2
(B) 3
(C) 4
(D) 6
(E) 7

5. If $x^2 - 7 = 9$, which of the following is a possible value for x^3?

(A) -64
(B) -16
(C) -8
(D) 8
(E) 16

6. Amy went to the grocery store and bought three equally-priced loaves of bread and three equally-priced containers of cold cuts. Amy spent a total of $18.00 and the price of the bread was k dollars per loaf and the price of cold cuts was $4.00 per container. How much do 6 loaves of bread cost?

(A) $6.00
(B) $9.00
(C) $12.00
(D) $18.00
(E) $24.00

7. If x is equivalent to -1/4 and y is equivalent to 16, which of the following has the least value?

(A) x + y
(B) x - y
(C) -xy
(D) x/y
(E) y/x

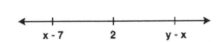

8. On the number line above, the tick marks are evenly spaced. What is the value of y^2?

(A)   9
(B)   11
(C)   25
(D)   49
(E) 121

GO ON TO NEXT PAGE

184

9. If $a + 2b + 3c = 12$ and $3a + 2b + c = 4$, what is the value of $a + b + c$?

   (A) 2
   (B) 4
   (C) 6
   (D) 8
   (E) 10

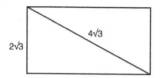

10. What is the area of the rectangle shown above?

   (A) $4\sqrt{3}$
   (B) 12
   (C) $12\sqrt{3}$
   (D) 24
   (E) $24\sqrt{3}$

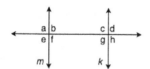

11. If line m is parallel to line k and h is less than 90^0, which of the following is NOT true?

   (A) $h + b = 180^0$
   (B) $g + a = 180^0$
   (C) $a + c < 180^0$
   (D) $a + f = 180^0$
   (E) $c + d + g > 270^0$

12. If $3x = 2y$, how many pairs of positive integers (x,y), in which both the x value and the y value are less than 10, satisfy the equation?

   (A) None
   (B) One
   (C) Two
   (D) Three
   (E) Four

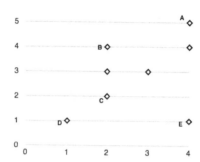

13. The scatterplot above shows the number of hours studied on Wednesday night (y-axis) versus the number of hours studied on Thursday night for 8 students. Which point represents the student who had the most drastic change in study hours between the two nights?

   (A) A
   (B) B
   (C) C
   (D) D
   (E) E

14. A ball is thrown upward from a height of 9 feet and the height of the ball, h, after t seconds can be modeled by the function $h(t) = 25 - (t - 4)^2$. How much time has passed by the second time the ball reaches a height of 21 feet?

   (A) 2
   (B) 4
   (C) 6
   (D) 8
   (E) 10

GO ON TO NEXT PAGE

A 1 1 1

Unauthorized copying or
reuse of any part of this
page is illegal.

1 1 1 **A**

15. The points R and S lie on a circle with center O. If the central angle, ∠ROS , measures 60⁰ and the perimeter of triangle ROS is 12, what is the area of the circle?

(A) 8π
(B) 16π
(C) 18π
(D) 24π
(E) 32π

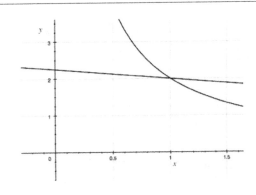

16. The functions f(x) and h(x) are shown in the xy-plane above. If f(b) = k and h(b) = k, what is the value of k?

(A) 0.5
(B) 1.0
(C) 1.5
(D) 2.0
(E) 2.5

17. In a school that offers two Physics courses, Physics 1 and AP Physics, there are 24 students who take AP Physics. If 25% of the students in the school take Physics and 25% of those students take Physics 1, how many students are in the entire school?

(A) 128
(B) 256
(C) 384
(D) 512
(E) 768

18. If $5^{2x+1} = 245$, what is the value of 5^x?

(A)   5
(B)   7
(C)   9
(D) 25
(E) 49

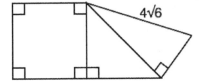

19. In the figure above, a 30-60-90 triangle rests against a 45-45-90 triangle, which in turn rests against a square. What is the area of the square if the hypotenuse of the 30-60-90 triangle measures 4√6?

(A)   8
(B) 12
(C) 24
(D) 36
(E) 72

20. The coordinate points (1,3) , (1,5) , and (4,6) define a triangle in the first quadrant of the xy-coordinate plane. If this triangle is reflected over the line y = x + 2, which of the following points lies on the resulting image?

(A) (1,3)
(B) (4,2)
(C) (6,4)
(D) (3,1)
(E) (5,1)

GO ON TO NEXT PAGE

A 2 2 2

Unauthorized copying or
reuse of any part of this
page is illegal.

2 2 2 A

SECTION 2
Time - 25 Minutes
18 Questions

Answer the questions in this section directly on the test or use a grid-in answer sheet.

Directions: This section contains two different types of questions and you have 25 minutes to complete both types. For questions 1-8, solve each problem and decide which is the best of the multiple-choice answers Fill in the corresponding circle in the answer bank or on a separate answer sheet. You may use the space available for scratch work.

Notes

1. Calculator usage is permitted.
2. All numbers utilized in the test are real numbers.
3. In this test, figures that accompany problems are there to provide information useful in solving the problems. The figures are drawn as accurately as possible EXCEPT when it is stated in a specific problem that the figure is not drawn to scale. All figures lie in a two-dimensional plane unless otherwise noted.
4. The domain of any function is assumed to be the set of all real numbers x for which the output of the function is a real number, unless specified otherwise.

Reference Information

   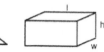

$A = \pi r^2$
$C = 2\pi r$ $A = lw$ $A = 1/2bh$ $V = lwh$ $V = \pi r^2 h$ $c^2 = a^2 + b^2$ Special Right Triangles

The total number of degrees of arc in a circle is 360.
The sum of the measures of the three angles of a triangle is 180 degrees.

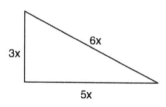

1. If the perimeter of the triangle above is 42, what is the value of x?

(A) 3
(B) 4
(C) 5
(D) 6
(E) 7

2. Suppose that the sum of x and y is 15. What is the average of x, y, and 15?

(A) 3
(B) 5
(C) 7.5
(D) 10
(E) 12.5

GO ON TO NEXT PAGE

3. What is the value of 3/7 of a number if five times the number is 140?

(A)   3
(B)   6
(C) 12
(D) 18
(E) 21

Type	# of Donuts
Vanilla	42
Chocolate	56
Glazed	28
Jelly	14

4. The table above shows the number of donuts that were sold at a donut shop on a certain morning. What percent of the donuts sold were vanilla donuts?

(A) 10
(B) 20
(C) 30
(D) 40
(E) 50

5. If x and y are both positive integers and $4y^2 = 64x$, which of the following must equal zero?

(A) $y + 4\sqrt{x}$
(B) $y - x$
(C) $4\sqrt{x} - y$
(D) $x^2 - y$
(E) $x^2 - y/4$

6. In the figure above, what is the area of the larger triangle on the right?

(A) $144\sqrt{3}$
(B) $108\sqrt{3}$
(C) $72\sqrt{3}$
(D) $54\sqrt{3}$
(E) $36\sqrt{3}$

7. ABC represents a three-digit integer in which A is divisible by 2, B is a perfect square, and C is the same as one of the other digits in the number. Which of the following could be the number?

(A) 246
(B) 642
(C) 862
(D) 888
(E) 898

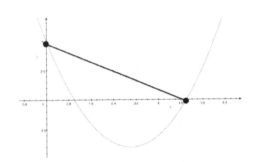

8. In the graph above, the function $f(x) = x^2 - 6x + 5$ intersects a line at the points shown. What is the slope of the line?

(A) -5
(B) -1
(C) -1/5
(D)   1
(E)   5

GO ON TO NEXT PAGE

A 2 2 2   2 2 2 A

Unauthorized copying or reuse of any part of this page is illegal.

Directions: For Student-Produced Response questions 9-18, feel free to write your answers directly below the questions. However, if you plan to use an answer sheet, use the grids at the bottom of the third page of the answer sheet.

The 10 questions that remain require you to solve the problems and enter your answers by marking the circles in the special grid, as shown in the examples below. You may use the space available for scratch work.

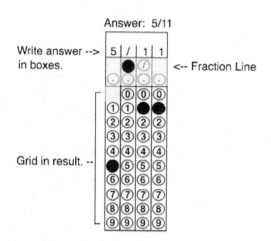

Answer: 5/11

Write answer --> in boxes.

<-- Fraction Line

Grid in result. --

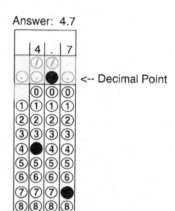

Answer: 4.7

<-- Decimal Point

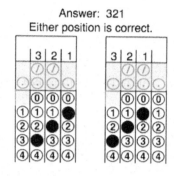

Answer: 321
Either position is correct.

Note: You may start your answers in any column, space permitting. Columns not needed should be left blank.

* Mark no more than one circle in any column.

* Because the answer sheet will be machine-scored, **you will receive credit only if the circles are filled in correctly.**

* Although not required, it is suggested that you write your answer in the boxes at the top of the columns to help you fill in the circles accurately.

* Some problems may have more than one correct answer. In such cases, grid only one answer.

* No question has a negative answer.

* **Mixed numbers** such as 4 1/2 must be gridded as 4.5 or 9/2. (If 4 1 / 2 is gridded, it will be interpreted as 41/2, not 4 and 1/2.)

* **Decimal Answers:** If you obtain a decimal answer with more digits than the grid can accommodate, it may be either rounded or truncated, but it must fill the entire grid. For example, if you obtain an answer such as 0.5555..., you should record your result as .555 or .556. **A less accurate value such as .55 and .56 will be scored as incorrect.**

Acceptable ways to grid 2/3 are:

9. In a toy chest, there are 4 trucks, 3 teddy bears, 5 robots, and 6 army men. If a child were to randomly select one toy from the chest, what is the probability that he will select a teddy bear or a robot?

$$67 < 3 + 16x < 99$$

10. What is one possible value of x that satisfies the inequality above?

GO ON TO NEXT PAGE

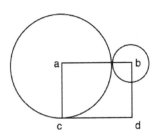

11. If the radius of circle *a* is 9 and the radius of circle *b* is 3, what is the area of the rectangle ABDC?

12. Suppose that $f(x) = (x - 4)^2 - 8$. If $k = 9$, what is the value of $f(k) / 17$?

13. If $(3.6 \times 10^4) / (1.2 \times 10^3) = d$, what is the value of d^2?

14. The degree measures of a quadrilateral are x^0, x^0, $(2x + 3)^0$, and $(x + 2)^0$. What is the measure of the largest angle?

15. A shipping company charges a flat fee of $4.95 to ship packages, plus $1.05 per pound. A competitor charges a flat fee of $0.75 and an additional fee of $1.25 per pound. What is the weight of a package that would cost the same to ship through either company?

16. A "Jimmy Cube" is any integer that is 5 less than twice a perfect cube. For example, 11 is a Jimmy Cube since 11 is 5 less than 16 which is twice 8; a perfect cube. What is the smallest Jimmy Cube that is greater than 100?

17. The volume of a cylinder is 108π and it has a height of 3. If a cylinder with a radius of 4 is removed from the larger cylinder, what is the volume of the original cylinder divided by the volume of the new modified cylinder?

18. An equilateral triangle is inscribed within a circle with an area of 12π. What is the perimeter of the triangle?

GO ON TO NEXT PAGE

SECTION 3
Time - 20 Minutes
16 Questions

Answer the questions in this section directly on the test or use a grid-in answer sheet.

Directions: For this section, solve each problem and decide which is the best of the multiple-choice answers. Fill in the corresponding circle in the answer bank or on a separate answer sheet. You may use the space available for scratch work.

Notes

1. Calculator usage is permitted.
2. All numbers utilized in the test are real numbers.
3. In this test, figures that accompany problems are there to provide information useful in solving the problems. The figures are drawn as accurately as possible EXCEPT when it is stated in a specific problem that the figure is not drawn to scale. All figures lie in a two-dimensional plane unless otherwise noted.
4. The domain of any function is assumed to be the set of all real numbers x for which the output of the function is a real number, unless specified otherwise.

Reference Information

  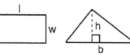   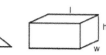

$A = \pi r^2$
$C = 2\pi r$ $A = lw$ $A = 1/2bh$ $V = lwh$ $V = \pi r^2 h$ $c^2 = a^2 + b^2$ Special Right Triangles

The total number of degrees of arc in a circle is 360.
The sum of the measures of the three angles of a triangle is 180 degrees.

---

1. If $3m = m + 10$ and $6d - 4m = 34$, what is the value of d?

(A) 8
(B) 9
(C) 10
(D) 11
(E) 12

2. If $24°$ cheesecake slices are cut from a circular cake, how many people can have a slice of cake?

(A) 12
(B) 15
(C) 18
(D) 20
(E) 24

GO ON TO NEXT PAGE

3. If $y = 2x + 3$, which of the following points is <u>not</u> a solution to the equation?

   (A) (1/2,4)
   (B) (-2,-1)
   (C) (-1,-1)
   (D) (1,5)
   (E) (2,7)

A, B, C, D, E, A, B, C, D, E, ...

4. If the sequence above repeats indefinitely, which of the following terms is a D?

   (A) 89
   (B) 90
   (C) 91
   (D) 92
   (E) 93

5. If four angles with consecutive angle measures are placed adjacent to one another, they form a right angle. What is the largest angle?

   (A) 20
   (B) 21
   (C) 22
   (D) 23
   (E) 24

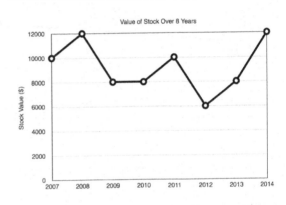

6. The line graph above shows the value of a certain stock over the course of 8 years. For how many of those years was the stock's value $8,000 at most?

   (A) 1
   (B) 3
   (C) 4
   (D) 6
   (E) 7

GO ON TO NEXT PAGE

7. If there are six farms on a huge plot of land and there are direct paths linking every farm to every other farm, how many paths are there?

(A) 21
(B) 15
(C) 10
(D) 6
(E) 3

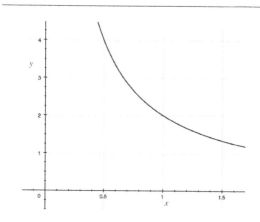

8. Which of the following statements about the graph of g(x), shown above, is false?

(A) f(0.5) > f(1)
(B) If f(a) = 2, then f(a) = 2a.
(C) If f(a) = 4, then f(a) = a/4.
(D) If f(a) > f(b), then a < b.
(E) As a increases, f(a) decreases.

9. Angela and David were running for class president. A total of 221 votes were cast and Angela won, having received 109 votes less than twice as many votes as David. How many votes did David lose by?

(A) 10
(B) 9
(C) 5
(D) 2
(E) 1

10. 14 consecutive integers sum to 119. What is the difference between the average of the numbers and the largest number in the group?

(A) 6
(B) 6.5
(C) 7
(D) 7.5
(E) 8

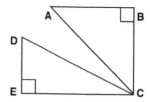

11. Suppose that EC ⊥ CB, triangle ABC is isosceles, DE measures 2, and EC measures 2√3. What is the measure of ∠ACD?

(A) 10^0
(B) 12^0
(C) 12.5^0
(D) 15^0
(E) 22.5^0

12. Derek is in charge of trying to make a more balanced workforce where he works. He works with people who are extremely sales-oriented and with people who are extremely customer-service-oriented. All of the employees are ranked on a scale from 1 to 20, 1 being completely sales-oriented and 20 being completely customer-service-oriented. Derek would like to select people who are more than 5 away from the perfect balanced score of 10 for a case study he is conducting. Which of the following represents all scores that are acceptable for his case study?

(A) | s - 10 | ≤ 5
(B) | s - 10 | < 5
(C) | s - 10 | ≥ 5
(D) | s - 10 | > 5
(E) | s - 5 | > 10

GO ON TO NEXT PAGE

13. x is the longest possible integer side length of a triangle with two other sides that measure 8 and 11. y is the shortest possible integer side length of a triangle with two other sides that measure 8 and 3. What is the value of x - y?

(A) 2
(B) 6
(C) 8
(D) 12
(E) 16

15. Daniel is making a bracelet with blue, red, yellow, and orange beads, which will be aligned in a repeating pattern that uses one bead of each color. The bracelet will require 87 beads to be completed. If Daniel is not sure which bead he will use to start the bracelet, how many total beads will he need to purchase?

(A) 87
(B) 88
(C) 90
(D) 92
(E) 96

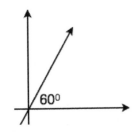

14. What is the slope of a line that is perpendicular to the line pictured above?

(A) -1/√3
(B) 1/√3
(C) -1/2
(D) 1/2
(E) -√3/1

16. The equation of line *g* is y = 3x - 2. Line *f* is perpendicular to line *g* and goes through the origin. If line *f* also goes through the point (3j, 4j - 2), what is the value of j?

(A) 2/5
(B) 3/5
(C) 3/2
(D) 5/3
(E) 5/2

STOP

Practice Test B

SECTION 1
Time - 25 Minutes
18 Questions

Answer the questions in this section directly on the test or use a grid-in answer sheet.

Directions: This section contains two different types of questions and you have 25 minutes to complete both types. For questions 1-8, solve each problem and decide which is the best of the multiple-choice answers Fill in the corresponding circle in the answer bank or on a separate answer sheet. You may use the space available for scratch work.

Notes

1. Calculator usage is permitted.
2. All numbers utilized in the test are real numbers.
3. In this test, figures that accompany problems are there to provide information useful in solving the problems.
   The figures are drawn as accurately as possible EXCEPT when it is stated in a specific problem that the figure is not drawn to scale. All figures lie in a two-dimensional plane unless otherwise noted.
4. The domain of any function is assumed to be the set of all real numbers x for which the output of the function is a real number, unless specified otherwise.

Reference Information

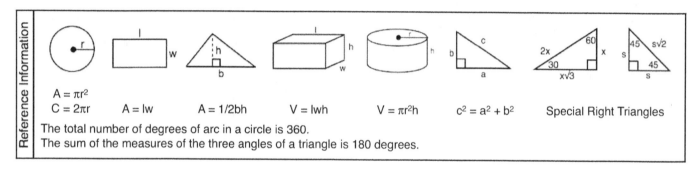

$A = \pi r^2$
$C = 2\pi r$ $A = lw$ $A = 1/2bh$ $V = lwh$ $V = \pi r^2 h$ $c^2 = a^2 + b^2$ Special Right Triangles

The total number of degrees of arc in a circle is 360.
The sum of the measures of the three angles of a triangle is 180 degrees.

---

1. Each month, a telephone service charges a base rate of $5.00 and an additional $0.04 per call for the first 30 calls, then $0.02 for every call after that. How much does the telephone service charge for a month in which 70 calls are made?

   (A) $2.00
   (B) $6.40
   (C) $7.00
   (D) $7.60
   (E) $7.80

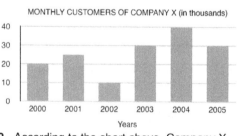

MONTHLY CUSTOMERS OF COMPANY X (in thousands)

Years

2. According to the chart above, Company X experienced its largest increase in monthly customers between which two consecutive years?

   (A) 2000 and 2001
   (B) 2001 and 2002
   (C) 2002 and 2003
   (D) 2003 and 2004
   (E) 2004 and 2005

GO ON TO NEXT PAGE

B 1 1 1

Unauthorized copying or
reuse of any part of this
page is illegal.

1 1 1 B

3. In the figure above, ∠PQS is 1/3 the measure of ∠PQR. If the measure of ∠PQR is 5/6 of a right angle, what is the measure of ∠PQS?

(A) 15^0
(B) 25^0
(C) 30^0
(D) 60^0
(E) 75^0

4. A retailer discounts a $30 item by 20%. If a customer purchases this item with an additional 15% off coupon, what is the final cost of the purchase?

(A) $9.60
(B) $10.50
(C) $19.50
(D) $20.40
(E) $25.50

5. If $(n + p)^2 < 1$, which of the following can be true?

I. $n > 1$
II. $|n + p| > 1$
III. $n = 2 - p$

(A) I only
(B) I and II
(C) II and III
(D) I and III
(E) None of the above

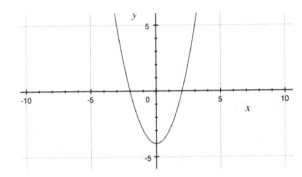

6. Which of the following could be the graph of the equation f, as shown above?

(A) $f(x) = x^2 - x - 4$
(B) $f(x) = x^2 + x - 4$
(C) $f(x) = x^2 - 4$
(D) $f(x) = x^2 + 4$
(E) $f(x) = x^2 - x + 4$

7. If s and t are positive integers, what percent of s is t + 10?

(A) $100(t + 10)/s$
(B) $100(t + 10)s$
(C) $100(s + 10)$
(D) $(t + 10)/s$
(E) $t/(s + 10)$

8. At noon, train X and train Y are heading toward one another along opposite tracks at rates of x and y miles per hour, respectively. If train Y travels 60 miles before it passes train X, how far apart, in miles, were the trains at noon?

(A) $60x + 60y$
(B) $x/60 + 60$
(C) $60x/y + 60/x$
(D) $60x/y + 60y$
(E) $60x/y + 60$

GO ON TO NEXT PAGE

Directions: For Student-Produced Response questions 9-18, feel free to write your answers directly below the questions. However, if you plan to use an answer sheet, use the grids at the bottom of the third page of the answer sheet.

The 10 questions that remain require you to solve the problems and enter your answers by marking the circles in the special grid, as shown in the examples below. You may use the space available for scratch work.

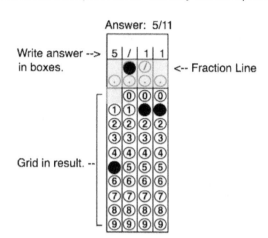

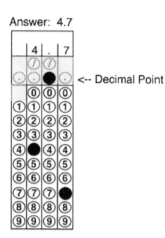

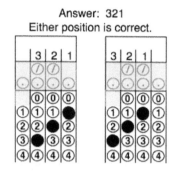

Note: You may start your answers in any column, space permitting. Columns not needed should be left blank.

* Mark no more than one circle in any column.

* Because the answer sheet will be machine-scored, **you will receive credit only if the circles are filled in correctly.**

* Although not required, it is suggested that you write your answer in the boxes at the top of the columns to help you fill in the circles accurately.

* Some problems may have more than one correct answer. In such cases, grid only one answer.

* No question has a negative answer.

* **Mixed numbers** such as 4 1/2 must be gridded as 4.5 or 9/2. (If 4 1 / 2 is gridded, it will be interpreted as 41/2, not 4 and 1/2.)

* **Decimal Answers:** If you obtain a decimal answer with more digits than the grid can accommodate, it may be either rounded or truncated, but it must fill the entire grid. For example, if you obtain an answer such as 0.5555..., you should record your result as .555 or .556. **A less accurate value such as .55 and .56 will be scored as incorrect.**

Acceptable ways to grid 2/3 are:

---

9. How many 1/2 pound weights put together weigh 14 pounds?

10. What is the value of n in the equation below?

$$\frac{(5 + n)}{4} = \frac{5}{2}$$

GO ON TO NEXT PAGE →

B 1 1 1

Unauthorized copying or
reuse of any part of this
page is illegal.

1 1 1 B

11. How many cubes, each with a side length of 2 inches, would it take to fill a rectangular prism with side lengths of 8 inches, 6 inches, and 2 inches?

12. If r is directly proportional to t and $t = 4$ when $r = 1/3$, what is t when $r = 6$?

13. A number, x, when divided by its cubed root is equal to 3 squared. What is the square of x?

14. A triangle with sides measuring 9, $3\sqrt{3}$, and $6\sqrt{3}$ has angles x, 3x, and y. What is the value, in degrees, of angle y?

Step 1: Choose an integer m.
Step 2: If m is odd, add 3; if m is even, multiply by 7.
Step 3: Take the ones digit of the result and square it.
Step 4: Divide the number by 4.

15. If the number m chosen in step 1 is 23, what number will be the result of step 4?

16. How many positive integers less than 200 are divisible by 5 or 7?

17. When John attempts to divide his Halloween candy into 5 piles, he finds he has 4 candies left over; when he tries to divide his candy into 4 piles, he has 3 candies left over; when he tries to divide his candy into 3 piles, he has 2 candies left over; and when he tries to divide his candy into 2 piles, he has one candy left over. What is the smallest number of candies John can have?

18. If x is an integer and $|(x^3 - x^2)| = 36$, what is the value of x^2?

GO ON TO NEXT PAGE

SECTION 2
Time - 25 Minutes
20 Questions

Answer the questions in this section directly on the test or use a grid-in answer sheet.

Directions: For this section, solve each problem and decide which is the best of the multiple-choice answers. Fill in the corresponding circle in the answer bank or on a separate answer sheet. You may use the space available for scratch work.

Notes

1. Calculator usage is permitted.
2. All numbers utilized in the test are real numbers.
3. In this test, figures that accompany problems are there to provide information useful in solving the problems. The figures are drawn as accurately as possible EXCEPT when it is stated in a specific problem that the figure is not drawn to scale. All figures lie in a two-dimensional plane unless otherwise noted.
4. The domain of any function is assumed to be the set of all real numbers x for which the output of the function is a real number, unless specified otherwise.

Reference Information

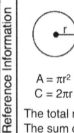  

$A = \pi r^2$
$C = 2\pi r$ $A = lw$ $A = 1/2bh$ $V = lwh$ $V = \pi r^2 h$ $c^2 = a^2 + b^2$ Special Right Triangles

The total number of degrees of arc in a circle is 360.
The sum of the measures of the three angles of a triangle is 180 degrees.

---

1. If $f + g = 17$ and $j + k = 34$, what is the value of $(f + g)/(j + k)$?

(A)   0
(B)   1/2
(C)   1
(D)   2
(E)   17

$$3 , 7 , 11 , x + 5 , 19 , ...$$

2. In the sequence above, the difference between any two consecutive terms is the same. What is the value of x?

(A)   10
(B)   13
(C)   14
(D)   15
(E)   17

GO ON TO NEXT PAGE ⟩

Weekday	Hours of Sleep
Monday	6
Tuesday	6
Wednesday	9
Thursday	7
Friday	8

3. The table above shows the hours of sleep for a certain student on the five weekdays of a single week. On average, how many hours of sleep did the student get during this week?

(A) 5.0
(B) 5.1
(C) 5.2
(D) 7.1
(E) 7.2

4. If the probability of getting an A on a test is 3/5 and the probability of getting a B is 2/9, what is the probability that a student will get an A on the first test, followed by B, on the second test?

(A) 1/15
(B) 2/15
(C) 5/14
(D) 2/3
(E) 37/45

5. If $9a^3 = -9$, what is the value of $f(a)$ if $f(x) = -2x + 3$?

(A) -15
(B) -1
(C) 1
(D) 5
(E) 21

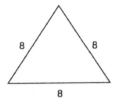

6. What is the area of the triangle pictured above?

(A) 32
(B) $16\sqrt{3}$
(C) 16
(D) $8\sqrt{3}$
(E) 8

7. Angela is going to lay out some holiday cards she received on her dining room table. She received five cards in the mail, but she only has room for three cards on the table. How many arrangements of three cards are possible?

(A) 12
(B) 24
(C) 60
(D) 72
(E) 120

8. Mary bought 5 apples and an orange at the market for $3.25. Bill bought 2 apples and an orange for $1.75. What is the cost of an orange?

(A) $0.25
(B) $0.50
(C) $0.75
(D) $1.67
(E) $2.50

GO ON TO NEXT PAGE

9. A $1,200 investment earns interest based on the following equation: $P(t) = 1200(1.04)^{m/2}$, where P is the investment's value after the number of months, m, it has been invested. What is the investment worth after 6 months have passed?

(A) $1,349.84
(B) $1,403.83
(C) $1,459.98
(D) $1,518.38
(E) $1,642.28

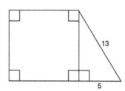

10. What is the area of the square pictured above?

(A) 100
(B) 121
(C) 144
(D) 169
(E) 196

11. Two vertical angles measure $(3x - 45)^0$ and $(x + 15)^0$. What is the degree measure of an angle that is $(3x)^0$?

(A) 15
(B) 30
(C) 45
(D) 60
(E) 90

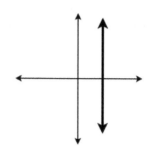

12. What is the equation of a line that is perpendicular to the line pictured above?

(A) $y = x + 1$
(B) $x = y + 1$
(C) $y = x$
(D) $y = 1$
(E) $x = 2$

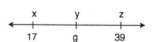

13. On the number line above, segment xy is equal in length to segment yz. What is the value of g?

(A) 26
(B) 27
(C) 28
(D) 29
(E) 30

$$y = -5x + 23$$

14. How many ordered pairs (x,y) of positive integers satisfy the equation above?

(A) None
(B) One
(C) Two
(D) Three
(E) Four

GO ON TO NEXT PAGE

B 2 2 2

Unauthorized copying or reuse of any part of this page is illegal.

2 2 2 B

15. How many positive integer values of x are solutions to the inequality $2x^2 - 4x - 96 < 0$

(A) 7
(B) 8
(C) 12
(D) 13
(E) 15

16. The average of x, y, and z is 24. The average of x, y, z, and p is 36. What is the value of p?

(A) 30
(B) 36
(C) 72
(D) 120
(E) 144

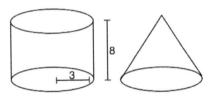

17. The cylinder and the cone pictured above have the same height and the same base radius. If the cone is filled with water and the water is then displaced into the cylinder, what will be the height of the water in the cylinder?

(A) 2/3
(B) 4/3
(C) 2
(D) 8/3
(E) 4

	9	10	11	12	TOTAL
Girls	28		18	22	80
Boys		32		40	
TOTAL			24		200

18. The incomplete table shown above displays the number of students in each grade level who are members of a certain club. How many students in the club are in the ninth or the tenth grade?

(A) 12
(B) 44
(C) 70
(D) 86
(E) 114

19. If $x^2 + y^2 = 124$ and $x - y = 10$, what is the value 2xy?

(A) 12
(B) 24
(C) 100
(D) 114
(E) 224

20. If a die with sides numbered one through six is rolled twice in a row, what is the probability that the sum of the two numbers rolled is greater than 9?

(A) 1/18
(B) 1/6
(C) 1/3
(D) 2/3
(E) 5/6

GO ON TO NEXT PAGE

SECTION 3
Time - 20 Minutes
16 Questions

Answer the questions in this section directly on the test or use a grid-in answer sheet.

Directions: For this section, solve each problem and decide which is the best of the multiple-choice answers. Fill in the corresponding circle in the answer bank or on a separate answer sheet. You may use the space available for scratch work.

Notes

1. Calculator usage is permitted.
2. All numbers utilized in the test are real numbers.
3. In this test, figures that accompany problems are there to provide information useful in solving the problems. The figures are drawn as accurately as possible EXCEPT when it is stated in a specific problem that the figure is not drawn to scale. All figures lie in a two-dimensional plane unless otherwise noted.
4. The domain of any function is assumed to be the set of all real numbers x for which the output of the function is a real number, unless specified otherwise.

Reference Information

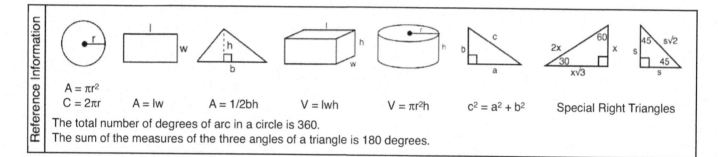

$A = \pi r^2$
$C = 2\pi r$ $A = lw$ $A = 1/2bh$ $V = lwh$ $V = \pi r^2 h$ $c^2 = a^2 + b^2$ Special Right Triangles

The total number of degrees of arc in a circle is 360.
The sum of the measures of the three angles of a triangle is 180 degrees.

1. If $x + 4x + 9x = 7$, then $x =$

(A) -1/2
(B) 0
(C) 1/2
(D) 2
(E) 14

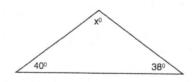

2. In the triangle above, $x =$

(A) 99
(B) 100
(C) 101
(D) 102
(E) 103

GO ON TO NEXT PAGE

3. For every 200 cameras manufactured by a factory, 8 are defective. At this rate, how many defective cameras, approximately, would be expected in a batch of 325 cameras?

(A) 13
(B) 21
(C) 24
(D) 32
(E) 80

4. If $15^8 = (15^2)(15^n)$, what is the value of n?

(A) 1/4
(B) 4
(C) 6
(D) 10
(E) 16

5. If 8 is y percent of 16, what is y percent of 30?

(A) 50
(B) 16
(C) 15
(D) 8
(E) 4

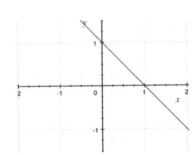

6. Line *l* is drawn such that it intersects (0,0) and is perpendicular to the line shown above. Which of the following points lies on line *l*?

(A) (-1,-1)
(B) (1,0)
(C) (0,1)
(D) (-1,1)
(E) (1,-1)

GO ON TO NEXT PAGE

7. If $3^x = 27$ and $x = 2y$, what is the value of $y + x$?

   (A) 1.5
   (B) 3
   (C) 4.5
   (D) 6
   (E) 9

8. If a and b are constants, which of the following is equal to the midpoint of the solutions to the equation $0 = 2(x - a)(x + b)$?

   (A) $(b - a)/2$
   (B) $(a - b)/2$
   (C) $(a + b)/2$
   (D) $2(b - a)$
   (E) $2(a - b)$

9. Which of the following is the closest approximation of the area of a circle whose circumference measures 2?

   (A) 0.101
   (B) 0.318
   (C) 0.405
   (D) 1.270
   (E) 3.143

10. If set A contains set R and set R contains s and t, which of the following must be true?

   (A) Set A contains s.
   (B) Set R contains set A.
   (C) When set A contains s, it contains t.
   (D) When set A does not contain s, it does not contain r.
   (E) When set A does not contain r, it contains both s and t.

11. If k is a constant, which of the following cannot represent the relationship between the volume, v, and the weight, w, of a uniform solid object?

   (A) $v = kw$
   (B) $1/v = k/w$
   (C) $v = w/k$
   (D) $v = k/w$
   (E) $v/w = k$

12. Which of the following is neither a prime number nor a composite number? (A composite number is defined as any integer that has at least one positive divisor other than one and the number itself.)

   (A) 1
   (B) 2
   (C) 3
   (D) 4
   (E) 10

GO ON TO NEXT PAGE ▷

$-a$, b , $2a$

13. If the average (arithmetic mean) of the three numbers above is b, what is b in terms of a?

(A) $2a$
(B) a
(C) $a/2$
(D) $-a/2$
(E) $-2a$

14. Let f be defined by $f(x) = 3x + 4$. If $f(a) = 38$, what is the value of a?

(A) 19
(B) 34/3
(C) 5
(D) 4
(E) 3

x , $3x + 1$, $2x + 5$

15. In the increasing sequence above, the difference between any two consecutive terms is constant. What is the value of the fourth term?

(A) 1
(B) 3
(C) 7
(D) 10
(E) Cannot be determined.

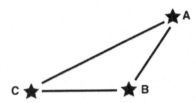

16. In the figure above, a signal is sent from point A to point C and will travel along both paths until terminated. Each time the signal traverses a line, it has a 1/3 chance of being terminated. How much greater is the probability of arrival at C when the signal is sent directly to C rather than sent through B to C?

(A) 1/27
(B) 2/9
(C) 1/3
(D) 8/27
(E) 4/9

STOP

Practice Test C

SECTION 1
Time - 25 Minutes
20 Questions

Answer the questions in this section directly on the test or use a grid-in answer sheet.

Directions: For this section, solve each problem and decide which is the best of the multiple-choice answers. Fill in the corresponding circle in the answer bank or on a separate answer sheet. You may use the space available for scratch work.

Notes

1. Calculator usage is permitted.
2. All numbers utilized in the test are real numbers.
3. In this test, figures that accompany problems are there to provide information useful in solving the problems. The figures are drawn as accurately as possible EXCEPT when it is stated in a specific problem that the figure is not drawn to scale. All figures lie in a two-dimensional plane unless otherwise noted.
4. The domain of any function is assumed to be the set of all real numbers x for which the output of the function is a real number, unless specified otherwise.

Reference Information

$A = \pi r^2$
$C = 2\pi r$ $A = lw$ $A = 1/2bh$ $V = lwh$ $V = \pi r^2 h$ $c^2 = a^2 + b^2$ Special Right Triangles

The total number of degrees of arc in a circle is 360.
The sum of the measures of the three angles of a triangle is 180 degrees.

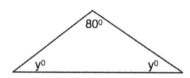

1. In the triangle above, what is the measure of y?

(A) 40
(B) 45
(C) 50
(D) 55
(E) 60

2. If 8.23 is rounded to the nearest tenth and the result is halved, what is the final result?

(A) 4.0
(B) 4.1
(C) 4.2
(D) 4.3
(E) 4.4

GO ON TO NEXT PAGE ⟩

3. Abby, Beth, and Charlotte want to compare their heights. Abby notices that Beth is taller than Charlotte, and Beth notices that Charlotte is shorter than Abby. Which of the following must be true?

(A) Beth is the same height as Abby.
(B) Beth is taller than Abby.
(C) Beth is shorter than Abby.
(D) Both Abby and Beth are taller than Charlotte.
(E) Charlotte and Beth are taller than Abby.

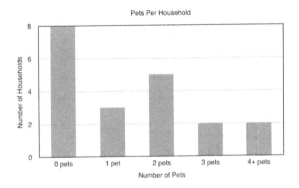

Pets Per Household

4. The graph above shows the results of a poll, which lists the number of pets each household in a neighborhood owns. How many houses were polled?

(A) 12
(B) 18
(C) 20
(D) 27
(E) 30

5. If 5 less than y is 3 more than 8, what is the value of y?

(A) -6
(B) -2
(C) 6
(D) 11
(E) 16

6. Which of the following equations has the property that the x-value and the y-value always add to the same number?

(A) y = -x
(B) y = x
(C) y = x + 2
(D) y = -2x
(E) y = -2x + 2

7. The rate of a particular chemical reaction is given by the equation R = (x - 3)(x + 20) + 60, where x grams is the mass of one of the chemical reactants. By how much does the rate increase if the mass of that reactant is increased from 14 to 24 grams?

(A) 10
(B) 434
(C) 550
(D) 700
(E) 760

8. The measures of two of the acute angles of a right triangle are 3z degrees and z degrees. How many degrees larger is the largest angle than the smallest angle?

(A) 22.5
(B) 30
(C) 45
(D) 67.5
(E) 90

GO ON TO NEXT PAGE

9. A student averages 84% on 5 tests and receives a perfect score of 100% on his final 3 tests. What is the student's average score for all 8 tests?

(A) 84%
(B) 86%
(C) 88%
(D) 90%
(E) 92%

10. A multiple of 10 between 10 and 110 (inclusive) is selected at random. What is the probability that the number will be divisible by 3?

(A) 3/11
(B) 4/11
(C) 6/11
(D) 3/10
(E) 6/10

11. If $k = 1 + t$ and $r + 1 = t - 1$, what is 1 more than k in terms of r?

(A) r + 1
(B) r + 2
(C) r + 3
(D) r + 4
(E) r + 5

12. How many one-on-one games of chess will be played in total if each of 7 players plays every other player once?

(A) 42
(B) 28
(C) 21
(D) 13
(E) 6

13. Which of the following is an equation for the line in the xy-plane that is perpendicular to the line $y = -2x + 3$ and contains the point (4,3)?

(A) $y = 1/2x + 1$
(B) $y = -1/2x + 1$
(C) $y = 1/2x + 3$
(D) $y = -2x + 11$
(E) $y = 2x + 5$

14. If $x = 3 - a$ and $y = 3 + a$, what is a^2 in terms of xy?

(A) xy + 3
(B) xy - 3
(C) xy - 9
(D) 9 - xy
(E) 9 + xy

GO ON TO NEXT PAGE

15. If 12 is b percent of c, what is c percent of b?

(A) 1/12
(B) 1
(C) 12
(D) 12/100
(E) 1/8

16. Two identical ropes form the circumferences of two small circles. The ropes are then arranged end-to-end to create a larger circle with a circumference twice the size of one of the small circles. What is the ratio of the area enclosed by the large circle to the sum of the areas enclosed by the two smaller circles?

(A) 4:1
(B) 2:1
(C) 4:3
(D) 1:1
(E) 1:2

17. If $f(x) = 2x^2$, which of the following is equivalent to $f(b - a)$?

(A) $(2b)^2 - (2a)^2$
(B) $2b^2 - 2a^2$
(C) $2b^2 - 2ba + 2a^2$
(D) $2b^2 + 4ba - 2a^2$
(E) $2b^2 - 4ba + 2a^2$

18. If x, y, and z are positive and xy > yz, which of the following must be true?

(A) $xy + x > yz + z$
(B) $xy + x < yz + z$
(C) $xy < -zy$
(D) $-xz > -zy$
(E) $x > y > z$

19. A sack of potatoes weighs 200 pounds. If the sack contains small bags of potatoes with an average weight of 4 pounds each, which of the following equations can be used to determine how many 7-pound bags of potatoes must be added to the sack so that the average weight of a bag in the sack is 5 pounds?

(A) $(50 + x)/5 = 200 + 7x$
(B) $(200 + 11x)/5 = 50 + x$
(C) $200 + 11x = 5(50 + x)$
(D) $200 + 7x = 50(5 + x)$
(E) $200 + 7x = 5(50 + x)$

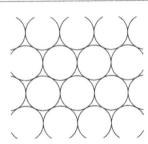

20. If the above pattern of tangent circles is repeated infinitely in all directions, what percent of the background is covered by the circles' areas, to the nearest decimal place?

(A) 47.6%
(B) 78.5%
(C) 88.6%
(D) 90.6%
(E) 95.2%

GO ON TO NEXT PAGE

SECTION 2
Time - 25 Minutes
18 Questions

Answer the questions in this section directly on the test or use a grid-in answer sheet.

Directions: This section contains two different types of questions and you have 25 minutes to complete both types. For questions 1-8, solve each problem and decide which is the best of the multiple-choice answers Fill in the corresponding circle in the answer bank or on a separate answer sheet. You may use the space available for scratch work.

Notes

1. Calculator usage is permitted.
2. All numbers utilized in the test are real numbers.
3. In this test, figures that accompany problems are there to provide information useful in solving the problems.
   The figures are drawn as accurately as possible EXCEPT when it is stated in a specific problem that the figure is not drawn to scale. All figures lie in a two-dimensional plane unless otherwise noted.
4. The domain of any function is assumed to be the set of all real numbers x for which the output of the function is a real number, unless specified otherwise.

Reference Information

  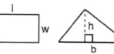  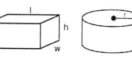

$A = \pi r^2$
$C = 2\pi r$ $A = lw$ $A = 1/2bh$ $V = lwh$ $V = \pi r^2 h$ $c^2 = a^2 + b^2$ Special Right Triangles

The total number of degrees of arc in a circle is 360.
The sum of the measures of the three angles of a triangle is 180 degrees.

1. If $4x + y = 24$ and $2x + 3 = 11$, what is the value of y?

(A) 3
(B) 4
(C) 6
(D) 8
(E) 16

2. 10 percent of 30 is 25 percent of what number?

(A) 15
(B) 12
(C) 4
(D) 3
(E) 1

GO ON TO NEXT PAGE

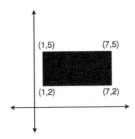

(1,5)   (7,5)

(1,2)   (7,2)

3. What is the area of the shaded quadrilateral above?

(A) 10
(B) 12
(C) 18
(D) 21
(E) 28

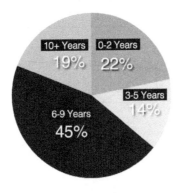

10+ Years
19%

0-2 Years
22%

3-5 Years
14%

6-9 Years
45%

4. The graph above shows the distribution of the number of years of work experience for a group of 5,000 adults. According to the graph, how many of these adults have 6 or more years of work experience?

(A) 1,100
(B) 2,250
(C) 2,800
(D) 3,200
(E) 3,900

5. If 1/(ab) = 1/k, which of the following is equal to 1?

(A) k
(B) ab
(C) (ab)/k
(D) (bk)/a
(E) abk

6. When the positive integer x is divided by 7, the remainder is 4. What is the remainder when 2x - 8 is divided by 7?

(A) 0
(B) 1
(C) 2
(D) 3
(E) 4

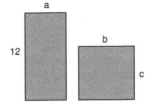

a

b

12

c

7. If the two rectangles above both have areas of 36 and c = 3a, which of the following must be true?

(A) b = 4a/3
(B) a = b
(C) a = 3b
(D) b = 12
(E) c = 3

8. All middleweight boxers in a competition must weigh between 166 and 184 pounds. If x pounds is the weight of a given boxer, which of the following expresses all acceptable values of x?

(A) |x - 175| = 9
(B) |x - 175| > 9
(C) |x + 175| > 9
(D) |x - 175| < 9
(E) |x + 175| < 9

GO ON TO NEXT PAGE

216

Directions: For Student-Produced Response questions 9-18, feel free to write your answers directly below the questions. However, if you plan to use an answer sheet, use the grids at the bottom of the third page of the answer sheet.

The 10 questions that remain require you to solve the problems and enter your answers by marking the circles in the special grid, as shown in the examples below. You may use the space available for scratch work.

Answer: 5/11

Write answer --> in boxes.

<-- Fraction Line

Grid in result. --

Answer: 4.7

<-- Decimal Point

Answer: 321
Either position is correct.

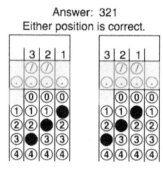

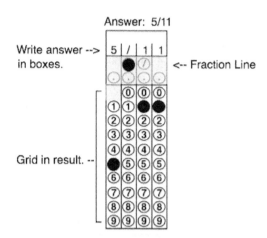

Note: You may start your answers in any column, space permitting. Columns not needed should be left blank.

* Mark no more than one circle in any column.

* Because the answer sheet will be machine-scored, **you will receive credit only if the circles are filled in correctly.**

* Although not required, it is suggested that you write your answer in the boxes at the top of the columns to help you fill in the circles accurately.

* Some problems may have more than one correct answer. In such cases, grid only one answer.

* No question has a negative answer.

* **Mixed numbers** such as 4 1/2 must be gridded as 4.5 or 9/2. (If 4 1/2 is gridded, it will be interpreted as 41/2, not 4 and 1/2.)

* **Decimal Answers:** If you obtain a decimal answer with more digits than the grid can accommodate, it may be either rounded or truncated, but it must fill the entire grid. For example, if you obtain an answer such as 0.5555..., you should record your result as .555 or .556. **A less accurate value such as .55 and .56 will be scored as incorrect.**

Acceptable ways to grid 2/3 are:

9. A class of 45 students is to be divided into teams of 2 or more players each. If each team must have the same number of students and every student must be on a team, what is the maximum number of teams possible?

10. What is the area of a right triangle whose sides have lengths 15, 20, and 25?

GO ON TO NEXT PAGE

796, 200, 51, 156, ...

11. In the sequence above, 796 is the first term, and every term thereafter is obtained by observing the following rules:

* If the previous term is even, add 4, then divide by 4.
* If the previous term is odd, multiply by 3, then add 3.
* If the final term is not an integer, the sequence finishes.

What is the final term of this sequence?

12. If 5a - 6b + a = 30, what is the value of 2a - 2b?

13. Jim played two rounds of an online video game in which six other players scored totals of 71, 66, 59, 92, 84, and 58. Jim's score in the first round was 41 and his total score, which is not included in the list above, was the only median amongst the seven scores. What is one possible score Jim could have earned in the second round?

14. The sum of a sequence of ten consecutive integers is -5. What is the largest integer?

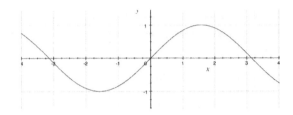

15. In the figure above, f(1/2) = b. For how many values of x between -4 and 4 does f(x) = b?

16. A spherical ball of clay with a radius of 9 inches weighs 5 pounds. How many pounds would a spherical ball of clay with a radius of 18 inches weigh? (Volume of a Sphere: $(4/3)\pi r^3$)

17. How many numbers between 1 and 1,000 have at least one digit equal to 5?

18. If $\sqrt{a}\sqrt{a^n} = \sqrt{3}$, what is $a^{(-2n-2)}$?

GO ON TO NEXT PAGE

218

SECTION 3
Time - 20 Minutes
16 Questions

Answer the questions in this section directly on the test or use a grid-in answer sheet.

Directions: For this section, solve each problem and decide which is the best of the multiple-choice answers. Fill in the corresponding circle in the answer bank or on a separate answer sheet. You may use the space available for scratch work.

Notes

1. Calculator usage is permitted.
2. All numbers utilized in the test are real numbers.
3. In this test, figures that accompany problems are there to provide information useful in solving the problems. The figures are drawn as accurately as possible EXCEPT when it is stated in a specific problem that the figure is not drawn to scale. All figures lie in a two-dimensional plane unless otherwise noted.
4. The domain of any function is assumed to be the set of all real numbers x for which the output of the function is a real number, unless specified otherwise.

Reference Information

  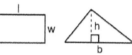  

$A = \pi r^2$
$C = 2\pi r$ $A = lw$ $A = 1/2bh$ $V = lwh$ $V = \pi r^2h$ $c^2 = a^2 + b^2$ Special Right Triangles

The total number of degrees of arc in a circle is 360.
The sum of the measures of the three angles of a triangle is 180 degrees.

---

1. What is the smallest value of $b + 8 > 15$?

(A) 5
(B) 6
(C) 7
(D) 8
(E) 9

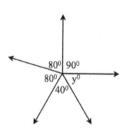

2. What is the value of y?

(A) 65
(B) 70
(C) 75
(D) 80
(E) 85

GO ON TO NEXT PAGE

---

3. At half-time, 2,400 people left their seats in a basketball stadium that seats 4,000. If 1,800 of those people returned when the second half of the game began, what fraction of the 2,400 did not return?

(A) 1/4
(B) 1/3
(C) 2/3
(D) 3/4
(E) 4/5

x	y
0.5	40
1.25	16
?	10
5	4

4. In the table above, x and y are inversely proportional to one another. What is the value of x when y is 10?

(A) 0.75
(B) 1
(C) 1.5
(D) 1.75
(E) 2

5. In a rectangular solid, four edges have lengths of 3, four edges have lengths of 4, and four edges have lengths of 6. What is the area of each of the largest faces on the solid?

(A) 12
(B) 18
(C) 24
(D) 28
(E) 36

6. A movie rental store charges p dollars for a VHS rental and $3p$ dollars for a DVD rental. If a customer pays $13p$ dollars for 5 movies, how many DVDs does he buy?

(A) Four
(B) Three
(C) Two
(D) One
(E) None

GO ON TO NEXT PAGE

7. If y is a multiple of three and $42 < 2y < 50$, what is $y + 5$?

(A) 26
(B) 27
(C) 28
(D) 29
(E) 30

---

8. Points P, Q, R, and S are adjacent vertices of a quadrilateral. Kate draws a new line through points P and R, and a second line through Q and S. She finds that the two lines intersect at a point outside of the quadrilateral. What shape must the quadrilateral be?

(A) Square
(B) Rectangle
(C) Trapezoid
(D) Parallelogram
(E) None of the above

---

9. Every time a broken 12-hour clock's hour hand reaches 1 o'clock, it freezes there for exactly one hour, then proceeds normally around the face of the clock. If James synchronizes the clock with his wristwatch at 12:00 noon on Monday, what hour is indicated by the broken clock's hour hand when James's wristwatch says it is 12:00 noon on Wednesday?

(A) 6 o'clock
(B) 7 o'clock
(C) 8 o'clock
(D) 9 o'clock
(E) 10 o'clock

$-2 , 1 , 4 , 7 , 10 , 13 , ...$

10. The first term of the sequence above is -2 and each term after the first term is three more than the preceding term. Which of the following expressions can be used to calculate the nth term of the sequence?

(A) $-5 + 3(n-1)$
(B) $-2 + (3n - 3)$
(C) $-2 + (3n - 1)$
(D) $-2 + 3n$
(E) $2 - 3n$

---

11. The function g is defined as $g(x) = ax^2 + bx - 6a$, where a and b are positive constants. The function has a line of symmetry at $x = -3$. Which of the following could NOT be a coordinate pair on the graph of $y = g(x)$?

(A) $(0,-3)$
(B) $(0,3)$
(C) $(-3,0)$
(D) $(3,3)$
(E) $(-3,-3)$

---

12. An athlete who jogs at a rate of x miles per hour begins to run along a straight road starting from her house. She plans to run $2x$ miles north, x miles south, $x/2$ miles north, and so on, until she tires out. How far, in miles, is the jogger from her home after running for 3.5 hours?

(A) 0
(B) 1
(C) 1.5
(D) 1.75
(E) 3.5

GO ON TO NEXT PAGE ⟩

C 3 3 3

Unauthorized copying or
reuse of any part of this
page is illegal.

3 3 3 C

13. Which of the following inequalities express only
values of x for which $x + x^{-1} > 2$?

    I. $x > 0$
    II. $|x - 1| > 0$
    III. $x - 1 > 0$

    (A) I only
    (B) I and II
    (C) I and III
    (D) III only
    (E) II and III

14. If $9^4 / 3^x = 3^y$ for positive integers x and y, what
is the value of x + y?

    (A) 6
    (B) 8
    (C) 10
    (D) 11
    (E) 12

$$f(2x) = [f(x)]^2$$

15. Given the equation above, if f(1) = 3, what is f(2)?

    (A) 1
    (B) 3
    (C) 9
    (D) 27
    (E) 81

16. In the figure above, point S is the center of the
semicircle, making QR a diameter. Point P is
located on the semicircle. What is the measure
of angle QPR?

    (A) 30^0
    (B) 45^0
    (C) 60^0
    (D) 90^0
    (E) 100^0

STOP

Chapter 1

1.1	1.2	1.3	1.4	1.5	1.6	1.7.1	1.7.2
1. A	1. D	1. B	1. C	1. B	1. C	1. B	1. D
2. B	2. B	2. D	2. A	2. C	2. A	2. A	2. C
3. B	3. C	3. C	3. A	3. D		3. E	3. E
4. E	4. B	4. B	4. B	4. A		4. C	4. B
5. D	5. A	5. A	5. D			5. D	5. A
	6. B	6. E	6. B			6. D	6. A
	7. C		7. D				
			8. E				
			9. C				
			10. D				

Challenge	Review	C. Review
1. 1	1. C	
2. 25	2. D	
3. 2	3. E	
4. 120	4. B	
5. 2.71	5. 3.75	
6. 122	6. 60	
7. 675	7. 55	
8. 5/9	8. 42	
9. 6		
10. 10		

Chapter 2

2.1.1	2.1.2	2.2	2.3	2.4
1. D	1. D	1. A	1. E	1. E
2. C	2. B	2. B	2. C	2. B
3. B	3. A	3. E	3. A	3. C
4. A	4. C	4. C	4. B	4. D
5. B	5. E	5. E	5. D	5. C
6. E	6. C	6. B	6. D	6. C
	7. A	7. C	7. C	7. A
		8. D	8. B	
		9. D	9. A	

Challenge	Review	C. Review
1. 80	1. A	1. D
2. 105	2. B	2. B
3. 141	3. C	3. E
4. 50	4. E	4. E
5. 64	5. D	5. A
6. 7	6. 210	6. 91
7. .88	7. 50	7. 50
	8. 34	8. 82
	9. 170	9. 50
	10. 5120	10. 32

Chapter 3

3.1.1	3.1.2	3.1.3	3.2	3.3	3.4	3.5	3.6	3.7	3.8.1	3.8.2	3.8.3	3.8.4
1. D	1. B	1. D	1. C	1. C	1. D	1. D	1. A	1. A	1. C	1. C	1. D	1. B
2. B	2. E	2. B	2. C	2. A	2. C	2. B	2. C	2. B	2. D	2. A	2. A	2. A
3. A	3. C	3. C	3. A	3. E	3. E	3. A	3. D		3. A	3. B	3. D	3. C
4. C	4. D	4. A	4. D	4. D	4. B	4. E	4. B		4. B	4. B	4. D	4. A
	5. A	5. E	5. B			5. E	5. C		5. C	5. C	5. D	5. E
	6. C	6. A				6. E			6. C	6. B	6. A	6. C
	7. A								7. D	7. D	7. C	
									8. B	8. D		
										9. E		

Challenge	Review	C. Review
1. 2	1. C	1. E
2. 5/14	2. E	2. B
3. 617	3. C	3. A
4. 2000	4. B	4. D
5. 12	5. 3/4	5. 7
6. 28	6. 42	6. 18
	7. 130	7. 99
	8. 21	8. 2400
	9. 2	9. 68

Chapter 4

4.1	4.2	4.3	4.4.1	4.4.2	4.4.3	4.5.1	4.5.2	4.5.3	4.5.4	4.5.5	4.6.1	4.6.2	4.6.3	4.7	4.8	Challenge	Review	C. Review
1. C	1. B	1. C	1. B	1. A	1. E	1. D	1. B	1. C	1. C	1. 14	1. A	1. B	1. 12.5	1. C	1. B	1. 717	1. D	1. C
2. A	2. B	2. D	2. D	2. D	2. B	2. D	2. C	2. D	2. B	2. 7/8	2. E	2. E	2. 15	2. A	2. A	2. 4	2. A	2. E
3. B	3. E	3. A	3. C	3. B		3. B	3. A	3. C	3. E	3. 256			3. 1/2	3. E	3. B	3. 7	3. C	3. C
4. D	4. A	4. B	4. A	4. B		4. A	4. E	4. A	4. D	4. 264			4. 24	4. D		4. 1/2	4. B	4. A
5. B	5. A	5. E	5. E	5. B		5. B				5. 1/3			5. 5	5. C		5. 72	5. A	5. 1/3
6. C	6. D		6. A	6. E						6. 3			6. 1/5	6. C		6. 1	6. 50	6. 20
7. E	7. A		7. C							7. 2/3			7. 2/5	7. B		7. 1/4	7. 2	7. 1
	8. C		8. C							8. 125			8. 8/9	8. A		8. 17/4	8. 99	8. 15
	9. D									9. 2			9. 1	9. E			9. 67	9. 45
	10. E									10. 10			10. C				10. |w - 130| ≤ 40	10. |x - 5.75| ≤ .75

Chapter 5

5.1	5.2.1	5.2.2	5.2.3	5.3	5.4	Challenge	Review	C. Review
1. A	1. E	1. B	1. C	1. E	1. C	1. 91	1. A	1. B
2. B	2. B	2. A	2. D	2. D	2. D	2. 314	2. C	2. D
3. B	3. C		3. A	3. A	3. E	3. 72	3. B	3. A
4. E			4. B	4. A	4. B	4. 60	4. C	4. A
			5. D	5. C	5. A	5. 77.9	5. 10	5. 2
			6. B	6. B		6. 240CG/XH	6. 50	6. 900
			7. A			7. 50		7. 102
						8. 24		8. 1200
								9. 205

Chapter 6

6.1.1	6.1.2	6.2	6.3	6.4	6.5.2	6.5.3	6.5.4	Challenge	Review	C. Review
1. A	1. E	1. C	1. D	1. D	1. C	1. B	1 - 9. [Sketches]	1. 17	1. C	1. B
2. D	2. A	2. B	2. D	2. E			10. III.	2. 1/2	2. D	2. C
3. C	3. B	3. A	3. A	3. C				3. 32.4	3. E	3. A
4. C	4. C	4. A	4. C	4. A				4. 3/8	4. 18	4. B
5. A	5. D							5. 12	5. 3	5. 1.35
6. D	6. D							6. 4/3	6. 61/4	6. 1902
	7. E							7. 5.27	7. 12/5	7. 5
										8. 15

6.5.1

	y-axis	x-axis	y = x
1.	y = -2x + 5	y = -2x - 5	y = 1/2x - 5/2
	y = 1/3x + 6	y = 1/3x - 6	y = -3x + 18
	y = -3/2x - 7	y = -3/2x + 7	y = 2/3x + 14/3
	y = -6x - 3	y = -6x + 3	y = 1/6x + 1/2
	y = 3	y = -3	x = 3

2. x-int: (6,0)   y-int: (0,6)

3. C

Chapter 7

7.1	7.2	7.3	7.4	7.5	7.6	7.7
1. A [Sketches on the Graph]		1. A	1. D	1. C	1. E	1. A
2. C		2. D	2. A	2. E	2. D	2. E
3. E			3. B	3. C	3. C	3. C
			4. A		4. E	4. B
			5. C		5. C	5. C
						6. A

Challenge	Review	C. Review
1. 0	1. D	1. B
2. 1	2. B	2. C
3. 1	3. E	3. A
4. 11	4. C	4. C
5. 96	5. 50	5. E
6. 768	6. 1	6. 2
7. 8/27	7. 21	7. 5/2
		8. 12

Chapter 8

8.1	8.2.1	8.2.2	8.2.3	8.2.4	8.2 MP	8.3	8.4	8.5	8.6
1. E	1. E	1. A	1. D	1. C	1. C	1. B	1. C	1. 3/25	1. A
2. C	2. D	2. B	2. E	2. B	2. D	2. C	2. B	2. 1/2	2. A
3. B	3. A	3. B		3. A	3. A	3. D	3. A		3. C
		4. C			4. D	4. D	4. E		
		5. D							

Challenge	Review	C. Review
1. 14	1. C	1. B
2. 8	2. C	2. E
3. 4	3. E	3. C
4. 40/3	4. C	4. B
5. 6.83	5. 6	5. 1.6
6. 15.4	6. .828	6. 36
7. 1/3		

Chapter 9

9.1	9.2	9.3	9.4	9.5	9.6	9.7	9.8	9.9.1	9.9.2
	1. D	1. B	1. B	1. C	1. D	1. C	1. A	1. B	1. B
	2. C	2. A	2. B	2. A	2. B	2. A	2. B	2. D	2. C
		3. B	3. E	3. B	3. C	3. B		3. D	3. C
		4. D	4. C		4. A	4. E		4. C	4. A
		5. E			5. E				

Challenge	Review	C. Review
1. 36	1. D	1. B
2. 6	2. D	2. B
3. 732	3. E	3. C
4. 10	4. A	4. A
5. 170	5. 2400	5. 60
6. 3	6. 1/9	6. 72
7. 22/9	7. 15	7. 150
8. 6/5		8. 1/4

Chapter 10

10.1.1	10.1.2	10.1.3	10.2	10.3	10.4.1	10.4.2	10.4.3	10.4 MP	10.5	10.6		Challenge	Review	C. Review
1. C	1. D	1. C	1. A	1. B	1. D	1. B	1. C	1. 21	1. D	1. D		1. 2	1. A	1. D
2. A	2. E	2. D	2. D	2. A	2. C	2. D	2. A	2. 6	2. B	2. E		2. 1	2. B	2. E
3. B	3. C	3. C	3. A	3. B	3. E	3. D	3. B	3. 120	3. B	3. E		3. 7776	3. B	3. B
4. B	4. C			4. C	4. C		4. E	4. 8736	4. C	4. D		4. 1440	4. 18	4. D
5. E	5. A			5. A	5. B			5. 72	5. A	5. A		5. 1/68	5. 6	5. 2.5
6. D				6. B	6. B			6. 35		6. D		6. 3/49	6. 48	6. 3
					7. B			7. 6720				7. 51		7. 1
					8. E			8. 1512						8. 3/4

INTEGRATED EDUCATIONAL SERVICES

Answer Key

PRACTICE TEST ANSWERS - Correct Answers and Difficulty Levels

Test A

Section 1

#	COR. ANS.	DIFF. LEV.	#	COR. ANS.	DIFF. LEV.
1.	A	1	11.	D	3
2.	E	2	12.	D	2
3.	D	1	13.	E	3
4.	C	2	14.	C	4
5.	A	2	15.	B	3
6.	C	2	16.	D	3
7.	E	3	17.	A	4
8.	E	2	18.	B	3
9.	B	2	19.	D	5
10.	C	3	20.	A	4

Section 2

Multiple-Choice	COR. ANS.	DIFF. LEV.	Student-Produced Response Questions	COR. ANS.	DIFF. LEV.
1.	A	1	9.	4/9	2
2.	D	1	10.	$4 < x < 6$	2
3.	C	2	11.	108	2
4.	C	2	12.	1	3
5.	C	2	13.	900	3
6.	D	3	14.	145	3
7.	E	3	15.	21	4
8.	B	3	16.	123	4
			17.	9/5	4
			18.	18	5

Section 3

#	COR. ANS.	DIFF. LEV.	#	COR. ANS.	DIFF. LEV.
1.	B	1	9.	E	3
2.	B	1	10.	B	3
3.	C	1	11.	D	3
4.	A	1	12.	D	4
5.	E	2	13.	D	3
6.	C	1	14.	A	4
7.	B	2	15.	B	5
8.	C	3	16.	A	3

Test B

Section 1

Multiple-Choice	COR. ANS.	DIFF. LEV.	Student-Produced Response Questions	COR. ANS.	DIFF. LEV.
1.	C	1	9.	28	2
2.	C	2	10.	5	2
3.	B	2	11.	12	3
4.	D	1	12.	72	3
5.	A	2	13.	729	2
6.	C	3	14.	60	3
7.	A	3	15.	9	4
8.	E	4	16.	62	5
			17.	59	4
			18.	9	5

Section 2

#	COR. ANS.	DIFF. LEV.	#	COR. ANS.	DIFF. LEV.
1.	B	1	11.	E	3
2.	A	1	12.	D	2
3.	E	1	13.	C	3
4.	B	2	14.	E	4
5.	D	2	15.	A	3
6.	B	2	16.	C	3
7.	C	2	17.	D	4
8.	C	3	18.	E	4
9.	A	2	19.	B	5
10.	C	3	20.	B	5

Section 3

#	COR. ANS.	DIFF. LEV.	#	COR. ANS.	DIFF. LEV.
1.	C	1	9.	B	2
2.	D	1	10.	A	3
3.	A	1	11.	D	3
4.	C	2	12.	A	4
5.	C	2	13.	C	4
6.	A	2	14.	B	3
7.	C	2	15.	D	5
8.	B	3	16.	B	4

Test C

Section 1

#	COR. ANS.	DIFF. LEV.	#	COR. ANS.	DIFF. LEV.
1.	C	1	11.	D	3
2.	B	2	12.	C	2
3.	D	1	13.	A	3
4.	C	1	14.	D	4
5.	E	2	15.	C	3
6.	A	2	16.	B	3
7.	C	2	17.	E	4
8.	D	2	18.	A	4
9.	D	2	19.	E	4
10.	A	2	20.	D	5

Section 2

Multiple-Choice	COR. ANS.	DIFF. LEV.	Student-Produced Response Questions	COR. ANS.	DIFF. LEV.
1.	D	1	9.	15	2
2.	B	1	10.	150	2
3.	C	2	11.	7/2 or 3.5	3
4.	D	1	12.	10	2
5.	C	2	13.	26, 27, 28, 29	3
6.	A	2	14.	4	3
7.	A	3	15.	3	3
8.	D	4	16.	40	4
			17.	271	5
			18.	1/9	5

Section 3

#	COR. ANS.	DIFF. LEV.	#	COR. ANS.	DIFF. LEV.
1.	D	1	9.	C	3
2.	B	1	10.	B	3
3.	A	2	11.	B	3
4.	E	2	12.	C	3
5.	C	2	13.	D	4
6.	A	2	14.	B	4
7.	D	3	15.	C	4
8.	E	2	16.	D	5

IES2400 Formula Chart

Problem Type	Formula and Method / Notes
Percent Change	New = Old$(1 \pm x\%)$
Combined Percentage	$1 \pm x\% = (1 \pm a\%)(1 \pm b\%)...$
Variation	Direct: $y/x = k$ Indirect: $yx = k$
Parts to the Whole	(Total)$(A/(A+B))$:$(B/(A+B))$(Total)
Combinations and Permutations	nPr (order matters) nCr (order doesn't matter)
Probability	$P(x) =$ Favorable/Total
Remainder	$N = Dx + R$
Speed	$D = ST$ Avg. Speed $= 2(S_1 \times S_2) / (S_1 + S_2)$
Working Rates	$x/r_1 + x/r_2 = 1$
Functions	$y = f(x)$
Number Facts	1 is not prime; 0 is not pos., neg., or prime
Sets	Venn Diagrams/Matrix/Chart
Factoring	$(x + y)^2 = x^2 + y^2 + 2xy$ $(x - y)^2 = x^2 + y^2 - 2xy$ $(x + y)(x - y) = x^2 - y^2$
Transformations (Quadratics)	Standard: a is shape, c is the y-int. Vertex: $f(x - h) + k$, (h,k) is the vertex.
Number Line	$0 < x < 1$ --> $x^3 < x^2 < x$ $-1 < x < 0$ --> $x < x^3 < x^2$
Prime Numbers	2, 3, 5, 7, 11, 13, 17, 19, 23, 29, 31, 37, ...
Slope	$m = (y_2 - y_1)/(x_2 - x_1)$ $y = mx + b$
Absolute Value	$|x - b| = a$, reads "What numbers are a away from b?"
Angles	Sum of interior angles $= 180/(n-2)$ Single exterior angle $= 360/n$
Arithmetic Sequences	$A_n = A_1 + d(n - 1)$ Sum $= n(A_n + A_1)/2$
Geometric Sequences	$A_n = A_1(r)^{n-1}$
Odds and Evens	Odd x Odd = Odd Odd + Even = Odd
Triangles	Area of an Equilateral $= s^2\sqrt{3}/4$ Third Side Rule: $|a - b| < c < |a + b|$
Growth and Decay	$P_t = P_0(1 \pm x\%)^{t/k}$